Time and Mind II: Information Processing Perspectives

Professor Dr. HEDE HELFRICH
Department of Psychology
University of Hildesheim
Marienburger Platz 22
31134 Hildesheim
Germany
e-mail: helfrich@uni-hildesheim.de

Time and Mind II: Information Processing Perspectives

edited by
Hede Helfrich

Library of Congress Cataloging-in-Publication Data

is now available via the Library of Congress Marc Database under the

LC Control Number: 2003115780

National Library of Canada Cataloguing-in-Publication

International Symposium on Time and Mind (2nd : 2002 : University of Hildesheim)
Time and mind II: information processing perspectives / Hede Helfrich, editor.

Papers from the International Symposium on Time and Mind 02, held at the
University of Hildesheim, Sept. 2-4, 2002.
Includes bibliographical references and index.
ISBN 0-88937-281-0

1. Time--Psychological aspects--Congresses. 2. Human information processing--Congresses. I. Helfrich, Hede II. Title.

BF468.I58 2002 153.7'53 C2003-906035-7

PUBLISHING OFFICES
USA: Hogrefe & Huber Publishers, 875 Massachusetts Avenue, 7th Floor, Cambridge, MA 02139
Phone (866) 823-4726, Fax (617) 354-6875, E-mail info@hhpub.com
Europe: Hogrefe & Huber Publishers, Rohnsweg 25, D-37085 Göttingen, Germany, Phone +49 551 49609-0, Fax +49 551 49609-88, E-mail hh@hhpub.com

SALES & DISTRIBUTION
USA: Hogrefe & Huber Publishers, Customer Services Department, 30 Amberwood Parkway, Ashland, OH 44805,
Phone (800) 228-3749, Fax (419) 281-6883, E-mail custserv@hhpub.com
Europe: Hogrefe & Huber Publishers, Rohnsweg 25, D-37085 Göttingen, Germany, Phone +49 551 49609-0, Fax +49 551 49609-88, E-mail hh@hhpub.com

OTHER OFFICES
Canada: Hogrefe & Huber Publishers, 1543 Bayview Avenue, Toronto, Ontario M4G 3B5
Switzerland: Hogrefe & Huber Publishers, Länggass-Strasse 76, CH-3000 Bern 9

Hogrefe & Huber Publishers
Incorporated and registered in the State of Washington, USA, and in Göttingen, Lower Saxony, Germany

Printed and bound in Germany
ISBN 0-88937-281-0

Table of contents

List of contributors

Richard A. Block
Department of Psychology, Montana State University, Bozeman, MT 59717-3440, USA; e-mail: block@montana.edu

Michel Bonnet
Laboratoire de Neurobiologie de la Cognition, Centre National de la Recherche Scientifique and Université de Provence, CNRS-GLM, 31 Chemin Joseph Aiguier, F-13402 Marseilles Cedex 20, France; e-mail: bonnet@lnf.cnrs-mrs.fr

Jean Bullier
Faculté de Médecine, Centre National de la Recherche Scientifique and Université Paul Sabatier, 133 route de Narbonne, F-31062 Toulouse Cedex 4, France; e-mail: bullier@cerco.ups-tlse.fr

Borís Burle
Laboratoire de Neurobiologie de la Cognition, Centre National de la Recherche Scientifique and Université de Provence, CNRS-GLM, 31 Chemin Joseph Aiguier, F-13402 Marseilles Cedex 20, France; e-mail: burle@lnf.cnrs-mrs.fr

István Czigler
Department of Psychology, Hungarian Academy of Sciences, Victor Hugo u. 18-22, H-1132 Budapest, Hungary; e-mail: czigler@cogpsyphy.hu

Russell M. Church
Department of Psychology, Brown University, Box 1853, Providence, RI 02912, USA; e-mail: Russell_Church@brown.edu

Michel Dojat
Institut National de la Santé et de la Recherche Médicale, Université Joseph Fourier, Laboratoire U594—Neuroimagerie Fonctionnelle et Métabolique, Centre Hospitalier Universitaire, Pavillon B, Box 317, F-38043 Grenoble Cedex 9, France; e-mail: Michel.Dojat@ujf-grenoble.fr

Andreas K. Engel
Institut für Neurophysiologie und Pathophysiologie, Universitätsklinikum Hamburg-Eppendorf, Martinistr. 52, D-20246 Hamburg, Germany; e-mail: engel@kognitionswissenschaft.de

Hans-Georg Geissler
Institut für Allgemeine Psychologie I, Universität Leipzig, Seeburgstraße 14-20, D-04103 Leipzig, Germany; e-mail: hans-g.geissler@rz.uni-leipzig.de

Robert B. Glassman
Department of Psychology, Lake Forest College, Lake Forest, IL 60045-2399, USA; e-mail: glassman@lfc.edu

Simon Grondin
École de Psychologie, Université Laval, Québec GIK 7P4, Canada; e-mail: simon.grondin@psy.ulaval.ca

Paulo Guilhardi
Department of Psychology, Brown University, Box 1853, Providence, RI 02912, USA; e-mail: Paulo_Guilhardi@brown.edu

Hede Helfrich
Institut für Psychologie, Universität Hildesheim, Marienburger Platz 22, D-31141 Hildesheim, Germany; e-mail: helfrich@uni-hildesheim.de

János Horváth
Department of Psychology, Hungarian Academy of Sciences, Victor Hugo u. 18-22, H-1132 Budapest, Hungary; e-mail: horvath@cogpsyphy.hu

Richard Keen
Department of Psychology, Brown University, Box 1853, Providence, RI 02912, USA; e-mail: Richard_Keen@brown.edu

Kimberly Kirkpatrick
Department of Psychology, University of York, York YO10 5DD, UK; e-mail: k.kirkpatrick@psych.york.ac.uk

Raul Kompass
Institut für Allgemeine Psychologie I, Universität Leipzig , Seeburgstraße 14-20, D-04103 Leipzig, Germany; e-mail: kompass@rz.uni-leipzig.de

Florian Klapproth
Institut für Psychologie, Universität Hildesheim, Marienburger Platz 22, D-31141 Hildesheim, Germany; e-mail: klapprot@rz.uni-hildesheim.de

Françoise Macar
Laboratoire de Neurobiologie de la Cognition, Centre National de la Recherche Scientifique and Université de Provence, CNRS-GLM, 31 Chemin Joseph Aiguier, F-13402 Marseilles Cedex 20, France; e-mail: macar@lnf.cnrs-mrs.fr

Mika MacInnis
Department of Psychology, Brown University, Box 1853, Providence, RI 02912, USA; e-mail: Mika_MacInnis@brown.edu

Thomas Rammsayer
Georg-Elias-Müller Institut für Psychologie, Universität Göttingen, Goßlerstr. 14, D-37073 Göttingen, Germany; e-mail: trammsa@uni-goettingen.de

Christoph Segebarth
Institut National de la Santé et de la Recherche Médicale, Université Paul Sabatier, Laboratoire U594—Neuroimagerie Fonctionnelle et Métabolique, Centre Hospitalier Universitaire, Pavillon B, Box 217, F-38043 Grenoble Cedex 9, France; e-mail : Christoph.Segebarth@inserm-u438.ujf-grenoble.fr

Elyse Sussman
Albert Einstein College of Medicine, Rose F. Kennedy Center, 1410 Pelham Parkway South, Bronx, NY 1046, USA; e-mail: esussman@balrog.aecom.yu.edu

Simo Vanni
Brain Research Unit, Low Temperature Laboratory, Helsinki University of Technology, Box 2200, 02015-HUT, Finland; e-mail: vanni@neuro.hut.fi

Jan Warnking
Institut National de la Santé et de la Recherche Médicale, Université Joseph Fourier, Laboratoire U594—Neuroimagerie Fonctionnelle et Métabolique, Centre Hospitalier Universitaire, Pavillon B, Box 317, F-38043 Grenoble Cedex 9, France; e-mail: Jan.Warnking@ujf-grnoble.fr

John H. Wearden
Department of Psychology, University of Manchester, Oxford Road, M13 9PL Manchester, UK; e-mail: wearden@psy.man.ac.uk

István Winkler
Department of Psychology, Hungarian Academy of Sciences, Victor Hugo u. 18-22, H-1132 Budapest, Hungary; e-mail: czigler@cogpsyphy.hu

Hirooki Yabe
Department of Neuropsychiatry, Hirosaki University, Hirosaki, 036-8560 Japan; e-mail: yabe@cc.hirosaki-u.ac.jp

Preface

This book focuses on the significance of time in information processing and thus broadens the perspective of the first *Time and Mind* volume (Helfrich, 1996).* Two areas of research are distinguished. The first deals with time estimation and the temporal coordination abilities of humans and animals. The second area of research deals with the role of temporal properties in processes involving perception and memory. Thus, the following two main topics constitute the structure of the book: "time as an object of information processing" and "time as a constituent factor in information processing." The aim of the book is to investigate both topics and to show possible ways to integrate the two research fields.

The book is based on the symposium "Time and Mind 02" held at the University of Hildesheim from September 2nd–4th, 2002. The Deutsche Forschungsgemeinschaft (DFG), the Universitätsstiftung, and the University of Hildesheim supported the idea of the symposium and provided grants to organize it.

Part I: Time as an object of information processing

First of all, animals and humans have no sense organ specialized to perceive time. Nevertheless, we speak of "time perception" and do not rule out the possibility that perceptual systems are tuned to temporal information. The processing of temporal information shares some important characteristics with the processing of visual, auditory, and other perceptual information. The theoretical model mainly discussed in this context is that of an *internal clock*. The contributions by RUSSELL CHURCH and colleagues as well as by JOHN WEARDEN focus on how an internal clock may operate as a kind of proximal or inner stimulus in both humans and animals. Although the metaphor of the internal clock has been useful for describing and exploring both time estimation and time-dependent behavior, there are—as Wearden points out—some unresolved problems in accounting for all the data available. The chapter by Church et al. shows that even non-human animals are able to combine multiple sources of temporal information. Thus, the question arises, whether one simple clock is sufficient to explain this ability or whether multiple clocks are required instead. The challenge increases regarding human processing of temporal information. While in most of the experiments usually one single response is required, human life provides numerous situations that require several independent temporal adjustments simultaneously (see the chapter by Block). For example, a pianist has to coordinate at least three temporal parameters: the duration of the tones, the changing pitch, and the temporal pattern of the hand and finger movements. Either, the internal clock must be capable of flexible adjustments, or several concurrently operating internal clocks would be needed to account for this ability.

---

* Helfrich, H. (Ed.). (1996). Time and mind. Seattle, WA: Hogrefe & Huber Publishers

Not all researchers take the existence of one or more internal clocks for granted. The contribution by RICHARD BLOCK reveals considerable influences of non-temporal factors, such as the *context* of time judgments and the allocation of *attention*. These influences may not be parsimoniously integrated in an internal-clock framework. It is an open question as to whether these non-timing factors can be treated as "response bias" and behavioral noise to be eliminated from "pure time," or whether they must be considered intrinsic components of psychological time.

A further challenge for internal-clock models is the question of the specificity of *sensory modalities* and their relation to time. This issue is addressed by the contributions of SIMON GRONDIN and FLORIAN KLAPPROTH. They question whether time perception and time memory are based on a central device independent of sensory modalities, or whether they are an emergent property of the way events are organized within each sensory modality. As Grondin points out, modality-specific timing devices may be adapted to different environmental requirements: Acoustic information such as speech and music is intrinsically based on temporal structure and, therefore, requires finer temporal resolution than optical information for which spatial structure is of crucial importance. This does not rule out the existence of a central timing device, as demonstrated for memory by Klapproth. He suggests an association between modality-specific encoding and a central time memory.

THOMAS RAMMSAYER investigates the neurobiological substrates of temporal processing and thereby tries to support the assumption of two distinct timing mechanisms, one for *short* and one for *long* durations. Based on experimentally induced pharmacological variations in the neurotransmitter systems, he presents evidence that temporal processing of intervals in the range of seconds or more is *cognitively* mediated, whereas processing of brief durations below approximately one-half second relies on automatic *sensory* mechanisms most likely located at a subcortical level.

The neurobiological approach leads to the general question whether these neural structures and functions are specific for time processing or, instead, are common to all information processing. With these considerations, the focus shifts, as Borís Burle says, "from temporal information processing to temporal processing of information." This issue is the focus of the second part of the book.

Part II: Time as a constituent of information processing

Time is an object of information processing, but it is also a constituent of it. There are at least three reasons, why time is constitutive for information processing. First, temporal structures of distal stimuli are essential for the contents of cognition. Second, temporal patterns of brain activity seem to be closely linked to formation of and attention to objects and events. And, third, each cognitive process is constrained by time limits.

In order to identify objects and events and not to experience the world as chaos, temporal *continuity* as well as temporal *change* must exist. The identity of objects in cognition is ensured only when consecutive elements are temporally integrated but separated from concurrent elements. Moreover, signals arriving from different sensory channels at different times must be synchronized in the brain.

Direct measurement of brain activity using electroencephalography as well as neuroimaging techniques lead to new insights about the ways the tasks of integration and separation are managed within the auditory and the visual modality. ISTVÁN CZIGLER and his colleagues analyze the temporal characteristics of auditory event-synthesis by using the mismatch negativity component (MMN) of event-related brain potentials. They conclude that there is a *temporal unit* of 200 ms (approximately the duration of a spoken syllable) where auditory stimuli are automatically perceived as belonging to the same event. SIMO VANNI and his research group combined temporal resolution techniques (MEG and EEG) with spatial resolution techniques (fMRI) to determine the *temporal succession* of activation in the visual system. Their results suggest that the timing of different areas may best be described as a feedforward-feedback cycle. While Vanni et al. focus on the asynchronous onsets of activity in different brain areas, the concept of *temporal binding* presented by ANDREAS ENGEL emphasizes the transient synchronization of neuronal discharges. Physiological experiments support the idea that the synchrony of neuroelectric oscillations binds together neural activity. This activity is broadly distributed among brain systems, and it is evoked by the various attributes of a stimulus object or another cognitive item. The synchronized short-time activity leads to traces in short-term (working) memory and thus enables the mental representation of objects.

According to contemporary models, working memory extends for approximately two seconds. Within this time span, several different elements can be held present. To account for the simultaneous presence of different elements, a theory of binding by synchrony must describe the time-dependent characteristics of neural processing in order to explain how each single element can be identified reliably and can be discriminated from other elements. ROBERT GLASSMAN suggests that simultaneously operating frequencies share a *harmonic structure*, similar to octave bands in musical information. According to this assumption, the lowest and highest frequencies simultaneously present in working memory share a proportion of 1:2. This mechanism allows the identification of distinct elements in working memory by repetition of each single synchrony. At the same time, the existence of sufficiently different synchronies allows the discrimination of different elements.

Concerning the temporal constraints of information processing, HANS-GEORG GEISSLER and RAUL KOMPASS start with the observation that all perceptual processes share common mechanisms across modalities. The results from several experiments in the field of perception and decision-making lead to the assumption that these mechanisms can best be described in terms of temporal constraints of our brain functioning. It seems that the human brain does not process information continuously but in successive steps of distinct duration, so-called *quantums* of time.

Some researchers have hypothesized that the periodic oscillations unfold on a background of a fixed oscillatory timing mechanism (an internal clock) that produces quantums of time for conscious information processing. According to this model, all cognitive processes (including time estimation) depend on a common neural-beat mechanism, or temporal rhythm. One interesting question is whether there exists a relationship between the frequency of the internal clock and the frequency with which other cognitive processes operate. BORÍS BURLE and his colleagues propose an architecture model, the kernel of which consists of an internal oscillator pacing both the flow of information and the receptivity state of neurons.

By putting together the different perspectives expressed by the various contributors to the book, an important step on the way towards a unified view of psychological time has been achieved. It can help to bridge the gap between "temporal information processing" and "temporal processing of information," between time as an object and time as a constituent of information processing.

The book has benefited from the fact that the contributors read and reviewed each other's chapters. I would like to thank them for this collaborative effort. I also thank Florian Klapproth, Carola Lindner-Müller, and Ulrich Seidler-Brandler who invaluably supported and advised me in organizing the symposium and preparing the book. Special thanks go to my colleague and husband Erich Hölter, from whom I received indispensable help and support. He not only encouraged me, but also kept track of the ultimate form of the book.

Hildesheim, November 2003 Hede Helfrich

Part I:

Time as an object of information processing

Chapter 1: Simultaneous temporal processing*

RUSSELL M. CHURCH, PAULO GUILHARDI, RICHARD KEEN, MIKA MACINNIS, and KIMBERLY KIRKPATRICK

Abstract

There is considerable evidence that animals can time multiple intervals that occur separately or concurrently. Such simultaneous temporal processing occurs both in temporal discrimination procedures and in classical conditioning procedures. The first part of the chapter will consist of the review of the evidence for simultaneous temporal processing, and the conditions under which the different intervals have influences on each other. The second part of the chapter will be a brief description of two timing theories: Scalar Timing Theory and Packet Theory of Timing. Scalar Timing Theory consists of a pacemaker-switch-accumulator system that serves as a clock, a memory that consists of examples of previously reinforced intervals, and a decision process that involves a comparison of ratios to a criterion; the Packet Theory of Timing consists of a conditional expected time function that serves as a clock, a memory that consists of weighted sums of these values, and a probabilistic decision process that produces packets of responses. Both of these theories will be applied to an example of simultaneous temporal processing by rats, and will serve as the basis for some general comments about the selection and evaluation of quantitative theories of timing.

Introduction

Rats, pigeons, and other animals readily learn to make time discriminations in the range of seconds to minutes. Such interval timing is typically demonstrated with fixed-interval procedures, but can also be seen in temporal discrimination procedures in which animals are trained to produce one response following an interval of a short duration and another response following an interval of a long duration (for example, see Stubbs, 1968).

In a fixed-interval schedule of reinforcement, the first response of an animal following a fixed interval of time (such as 60 s) is followed by food. As a result of such training, animals readily learn to respond more rapidly late in the interval than early in the interval (for example, see Catania & Reynolds, 1968). In a standard operant fixed-interval schedule of reinforcement, the fixed interval is defined as the time

* Preparation of this chapter was supported by National Institute of Mental Health Grant MH44234 to Brown University.

from the delivery of food until the availability of the next food. Alternatively, if a stimulus precedes the food, the interval may be specified from the onset of a stimulus until the availability of the next food. Whether the time marker is the previous food or stimulus onset, the next food is delivered contingent upon the response.

Variations on the fixed-interval procedure have been undertaken to determine whether animals can simultaneously time multiple intervals at once. In a segmented fixed-interval procedure (described in further detail below) there are two potential time markers, the event that marks the beginning of the fixed interval, and the event that marks the beginning of a segment. It is possible that an animal can time both the fixed interval and the segment simultaneously. The segmented fixed-interval procedure and its major results will be discussed in the first portion of the chapter.

In a search of the PsychINFO database for the years between 1887 and 2002, only five articles were found in which the phrase "simultaneous temporal processing" appeared in the title or the abstract. The first of these was an article by Meck and Church (1984). The other four articles that used this phrase in the title or the abstract included either Meck or Church as one of the authors (Church, 1984; Meck, 1987; Olton, Wenk, Church, & Meck, 1988; Meck & Williams, 1997). Based on these facts, one might assume that there is little evidence for simultaneous temporal processing, but that would be mistaken. Many standard conditioning experiments contain multiple time markers that can be timed simultaneously. Although most of these procedures were not explicitly designed to produce simultaneous temporal processing, there is ample opportunity for such timing to occur. The second portion of the chapter will discuss the form of simultaneous temporal processing under widely implemented standard conditioning procedures.

The segmented fixed-interval procedure

One variant of the fixed-interval procedure is to add another stimulus during the interval. This procedure, which will be referred to as a "segmented fixed-interval procedure," has been used for at least three different purposes: as a test of the chaining hypothesis, as a test of the conditioned reinforcement hypothesis, and as a test of the simultaneous temporal processing hypothesis. Although all the experiments to be described used a comparison of a standard fixed-interval procedure with a segmented fixed-interval procedure, they differed in many ways. Procedures were used in which the fixed interval was specified from the delivery of the previous food, and also procedures in which the fixed interval was specified from the onset of a stimulus; in some procedures the onset of a segment was delivered at a fixed time while in other procedures it was delivered following the first response after a fixed time; both pigeons and rats were used; the duration of the intervals differed considerably. The fixed intervals ranged from 50 s to 60 min; the duration of the segment stimulus varied from 0.7 s to 50 s; and the duration of the segments varied from 10 s to 4 min. Despite these differences in durations and the differences in the interpretations of the results, the response gradients in the experiments with the segmented fixed-interval procedures were similar.

As a test of the chaining hypothesis of fixed-interval performance

In a fixed-interval procedure, the mean response rate of the animal increases as a function of time. Although it might be natural to assume that "time" was the independent variable, there has been an extensive and continuing effort to identify the observable, or at least potentially observable, responses that occur during the time interval that may serve as discriminative stimuli. Behavior in a fixed-interval procedure can be characterized as a series of responses, and the assumption is that reinforcement strengthens responses that occurred shortly before its delivery more strongly than responses that occurred earlier. The chaining hypothesis is that each link in the behavioral chain acts as a discriminative stimulus that controls the response rate during the next link. Thus, the series of responses may serve as a mediating behavior between the successive deliveries of food. This is known as the chaining hypothesis of fixed-interval performance (Keller & Schoenfeld, 1950). One test of the chaining hypothesis is to present a stimulus during the fixed time interval that disrupts performance during the stimulus. According to the chaining hypothesis, such a stimulus should also affect the overall increase in mean response rate as a function of time.

In one experiment, four pigeons were trained on a fixed-interval procedure in the first phase and a segmented fixed-interval procedure in the second phase (Dews, 1962). In the fixed-interval procedure, food was available 500 s after the previous food delivery. In the segmented fixed-interval procedure, the houselight was off for 50 s, on for 50 s, etc. throughout the 500-s interval from food until the availability of the next food. This segmented fixed-interval procedure is illustrated at the top of Figure 1.

The results of the experiment are also shown in Figure 1. The independent variable is time in seconds since the last delivery of food; the dependent variable is response rate as a proportion of the maximum response rate. During the first phase of fixed-interval training (open squares with dotted lines), the mean response rate increased as a function of time. This pattern is often referred to as a "scallop." During the second phase with segmented fixed-interval training, there was a marked decrease in response rate when the houselight was off. The use of a single measure of response rate during each segment obscures any gradient of responding within segments, but gradients following the onset of a segment stimulus can be seen in the results of the next two experiments to be described (Figures 2 and 3).

This reduction in response rate when the houselight was off could have occurred for many reasons. It may have been a disrupter (Pavlovian external inhibition); it may have been because food was never delivered when the houselight was off (Pavlovian discriminative inhibition); or it may have been due to the difference in salience of the presence or absence of the houselight. For the test of the chaining hypothesis, the cause of the reduction in response rate when the houselight was off relative to when the houselight was on was not important. The critical observation was that, during the segmented fixed-interval procedure, the mean response rate in the presence of the houselight increased as a function of time, i.e., the scalloped pattern remained. Apparently, the maintenance of this temporal gradient of responding did not require the maintenance of responding during the time that the houselight was off. This finding, coupled with the fact that the response rates during the terminal segments were approximately the same in the FI and the segmented FI conditions, was evidence against the chaining hypothesis. The temporal gradient could be maintained in the absence of mediating responses. A series of studies by Dews increased generality for these results and provided additional support for these conclusions (Dews, 1965a, 1965b, 1966, 1970).

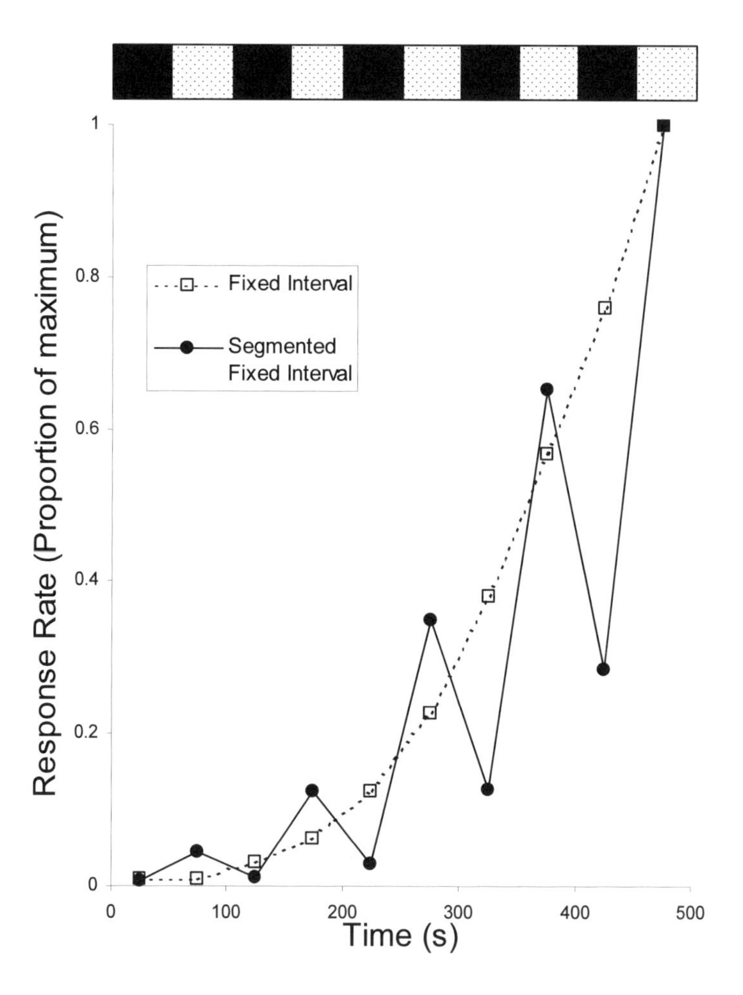

Figure 1: Relative response rate as a function of time since food in a fixed-interval procedure (open squares) and a segmented fixed-interval procedure (solid circles). The segmented fixed-interval procedure is illustrated at the top of the figure: A light was off during the dark intervals and on during the light intervals. Redrawn from Dews (1962).

As a test of the secondary reinforcement hypothesis

A primary reinforcer is normally defined as something that satisfies a biological need, such as hunger. Thus, food is a primary reinforcer. A conditioned reinforcer may be created by pairing a previously neutral stimulus with a primary reinforcer. In the segmented fixed-interval procedure, food is delivered in the presence of one of the segments. Thus, the segment, and others like it, should be conditioned reinforcers and would serve to reinforce behavior. One interpretation of the behavior of animals in a segmented fixed-interval procedure is that the behavior is maintained by conditioned reinforcement.

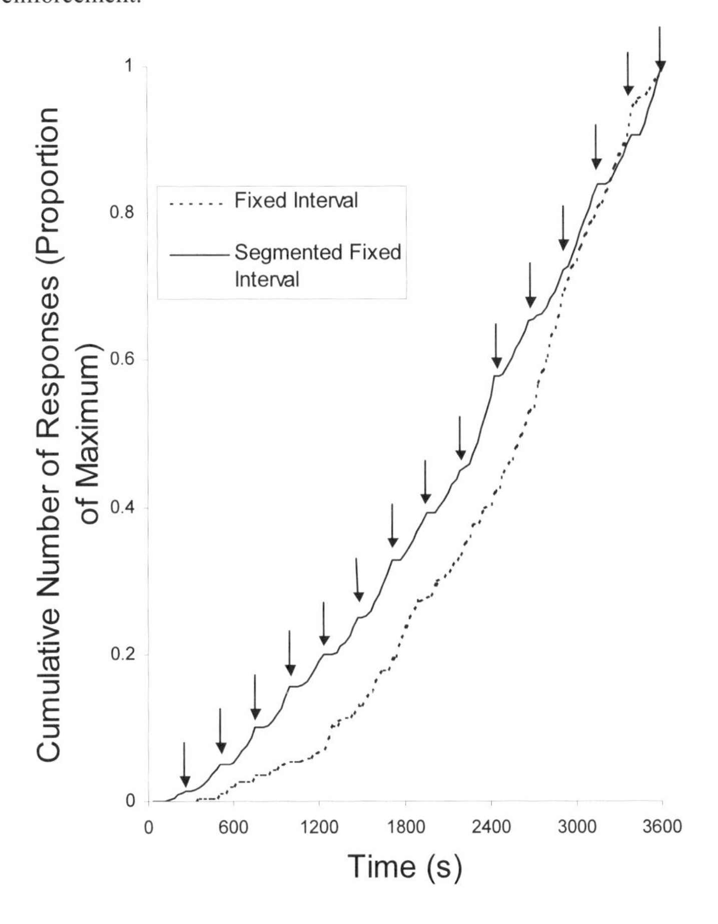

Figure 2: Relative cumulative number of responses as a function of time since food in a fixed-interval procedure (dashed line) and a segmented fixed-interval procedure (solid line). The arrows indicate the time at which a 0.7-s light stimulus was presented. Redrawn from Kelleher (1966).

Figure 2 shows the procedure and some results in one experiment with pigeons (Kelleher, 1966). The independent variable is the time since the last food delivery and the dependent variable is the relative cumulative number of responses (the number of responses during a small interval of time, divided by the total number of responses). The arrows indicate the times at which short stimuli (0.7-s lights) occurred. Each segment stimulus was delivered following the first response after a 4-min interval. The data are taken from only a single 60-min interval for the fixed-interval condition (the dashed line), and from the mean of only two 60-min intervals for the segmented fixed-interval condition (the solid line). The response rate is represented by the slope of the cumulative response function. There was a rising slope in the normalized cumulative response function for both the fixed-interval procedure and the segmented fixed-interval procedure. In addition, for the segmented fixed-interval procedure, response rate was low immediately after a segment stimulus, and then higher later in the 4-min interval.

The main purpose of the experiment was to determine whether or not the segment stimuli could increase relative response rate during a long (60-min) fixed-interval schedule of reinforcement, and whether or not they could lead to the development of a within-segment response gradient. Both of these findings were reported, and they were considered to be supportive of the conditioned reinforcement hypothesis. In other experiments in this article, the segment stimulus was not presented at the end of the last segment which was immediately before delivery of the reinforcer. In these experiments, the response gradients in the segments depended on the pairing of the segment stimulus with the food reinforcement. This supported the interpretation that the segment stimulus was a conditioned reinforcer. However, in a more thorough analysis of the determinants of conditioned reinforcement, Stubbs (1971) did not find a difference in performance between presentation of segments paired or not paired with food reinforcement, even when factors such as the animal's history, reinforcement schedule, and reinforcement rate were analyzed. This suggested that the segmented stimulus served as a discriminative stimulus rather than as a conditioned reinforcer.

As a test of the simultaneous temporal processing hypothesis

Meck and Church (1984) attempted to determine whether rats might simultaneously time the segments in conjunction with timing of the fixed intervals. In the first of the seven experiments the first phase consisted of 35 3-hour sessions of fixed-interval training followed by a second phase with 30 3-hour sessions of segmented fixed-interval training. For fixed-interval training, after an interval of 130 s in a dark box, a houselight was turned on. The first lever response after a fixed interval of 50 s delivered a 45-mg pellet of food and turned off the houselight. These cycles were repeated throughout the session. The segmented fixed-interval procedure was the same, except for the addition of 1-s white noise stimuli that occurred at the time of houselight onset, and 10, 20, 30, and 40 s after the time of houselight onset (as shown at the top of Figure 3).

Figure 3 shows the response rate (expressed as a percentage of the maximum response rate) on the last 20 sessions of fixed-interval training and on the last 20 sessions of segmented fixed-interval training. During the fixed-interval training, the mean response rate had the standard increasing gradient, the fixed-interval scallop. During the second phase with segmented fixed-interval training, there was an overall increase in response rate as a function of time since onset of the houselight, but also a clear decrease in response rate at the onset of the white-noise segment stimuli. In terms of relative response rate, the magnitude of the effect increased as the time of the next reinforcement approached.

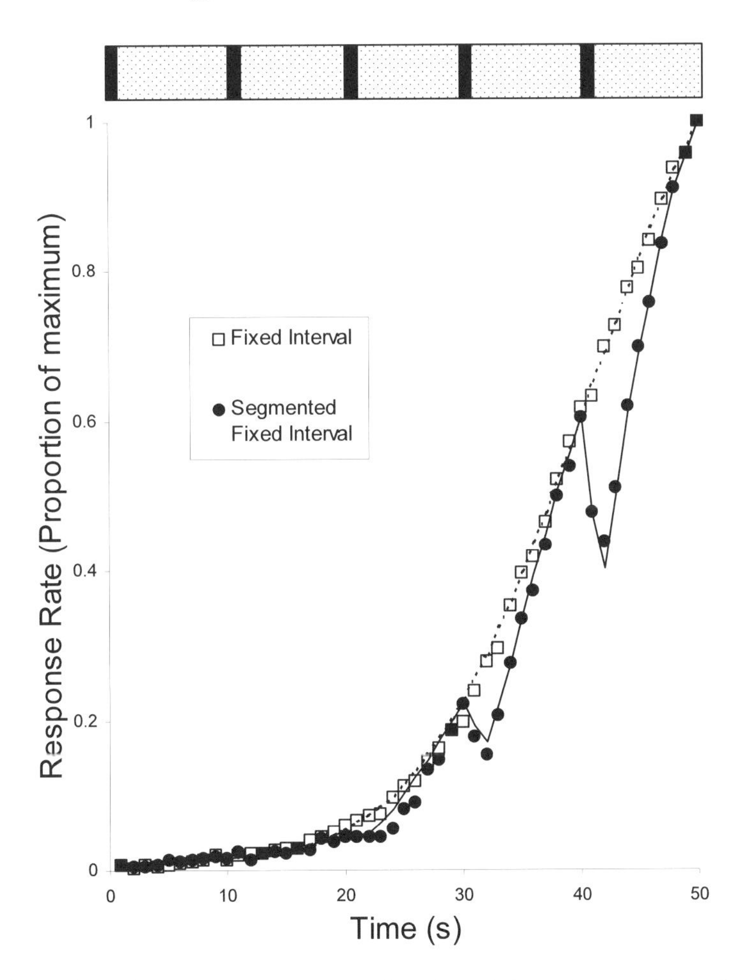

Figure 3: Relative response rate as a function of time since food in a fixed-interval procedure (open squares) and a segmented-fixed interval procedure (solid circles). The segmented fixed-interval procedure is illustrated at the top of the figure: The dark intervals indicate the times at which a 1.0-s white noise stimulus was presented. Redrawn from Meck and Church (1984).

In a second experiment, Meck and Church (1984) repeated the conditions shown in Figure 3, but added one more white-noise segment stimulus during the last second of the interval. The overall and segment gradients were similar to those shown in Figure 3. Other experiments in this article produced similar results with a segmented peak procedure in which the fixed interval and segment stimuli continued beyond the normal time of reinforcement (and reinforcement was withheld), and when the final segment stimulus occurred just before the reinforcer. These results suggest that the onset of the stimulus for the fixed interval, and the onset of a segment stimulus both served as discriminative stimuli for the time at which food would be available.

The main contributions of the Meck and Church (1984) experiments were to describe the problem as one of timing multiple intervals (rather than disruption of a response chain or conditioned reinforcement), and to describe the results of a quantitative model of timing. With this timing perspective it became natural to examine whether the application of scalar timing theory to a single interval could be extended to the timing of multiple intervals. In Figures 1 and 2 the lines merely connected the observed data points, thus facilitating visualization of the pattern of the data. In Figure 3 the dotted and solid lines that were near the observed data points were based on a quantitative theory of timing—scalar timing theory. It is possible that an examination of scalar timing theory will provide some understanding of the basis of simultaneous temporal processing.

Scalar timing theory

The essential principles of scalar timing theory were developed by Gibbon (1977), and they were subsequently used in the development of an information processing model of scalar timing theory (Gibbon, Church, & Meck, 1984). The information-processing model of scalar timing theory contains three major parts: clock, memory, and decision. A clock consisted of a pacemaker, a switch, and an accumulator; the memory was a reference memory for long-term storage of time intervals, and the decision was done by a comparator that could measure the nearness of the current time (in the accumulator) with a remembered time that was sampled from a reference memory. For timing a single interval, all that is needed is a single clock (pacemaker, switch, and accumulator), a single memory, and a single comparator. These parts are shown in the upper left side of Figure 4.

In the fixed-interval procedure mentioned above, when the food was delivered the food onset switch would close, permitting pulses from the pacemaker to enter the accumulator. Thus, if the pacemaker emitted 5 pulses a second with no variability, after 50 s the accumulator would have 250 pulses. If there was some variability in the pacemaker rate, after 50 s the accumulator might have fewer or more than 250 pulses. Reference memory contained a representation of the number of pulses in the accumulator at times that reinforcement had been received in the past. This is an exemplar memory that contains separate representations for each of the past examples. The decision is based on a comparison of the current accumulator value, which is continually increasing, and the value of a random sample of one element from reference memory. The comparator output depends on a ratio comparison of the two inputs (accumulator and memory) and a threshold criterion. If the current accumulator value is close enough to the value of the sample from memory, a response occurs. Details of this model are described in several sources (Church & Gibbon, 1982; Gibbon, Church, & Meck, 1984; Church, 2003). Four sources of variability were implemented in the simulations: the coefficient of variation (ratio of standard deviation to mean) of the clock, the mean and standard deviation of the threshold criterion, and the probability of inattention. An exhaustive search of the parameter space produced optimal values of the parameters of .25, .20, .10, and .01 for the four parameters, and these same parameter values were used in several experiments. This accounted for over 99% of the total variance in the response rate gradients in the fixed-interval condition, which suggests that, if scalar timing theory is correct, the animals were nearly always attentive to the time, but that there was some clock and threshold variability.

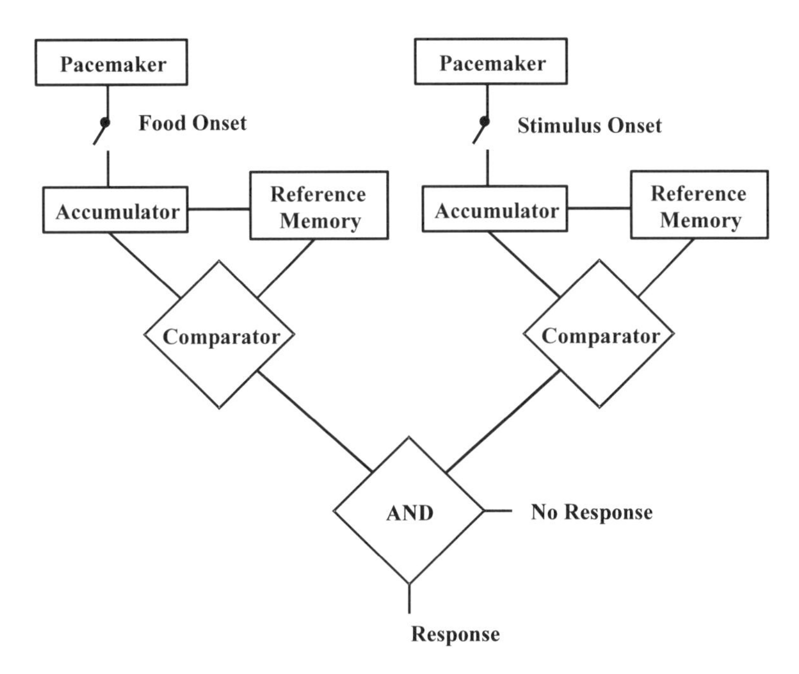

Figure 4: Application of scalar timing theory to the segmented fixed-interval procedure. Two independent clock-memory-decision units are controlled by food onset and stimulus onset. The output of these two units produces a response if both are above a threshold.

In the segmented fixed-interval procedure, the same process with the same parameters was used for the time since food onset, and an equivalent process (but with the addition of a latency-to-close-the-switch parameter, and different parameter values) was used for the time since stimulus onset (the top right part of Figure 4). Thus there was simultaneous timing of the interval since food and the interval since stimulus onset. The output of the two comparators were combined by assuming that the animal attended to the overall interval with some probability, that it attended to the segment interval with some probability, and that these two probabilities were independent. Thus, on some occasions, the rat attended to both intervals, on some it attended to only the overall interval, and on some it attended to neither. This accounted for over 99% of the total variance in the response rate gradients in the segmented fixed-interval condition, which suggests that, if scalar timing theory is correct, the animals were combining information from the overall and segment intervals in the determination of whether or not to respond.

Simultaneous temporal processing in conditioning procedures

The performance of animals in the segmented fixed-interval procedure makes it clear that they are able to time two intervals simultaneously. Is this an ability that requires a particular test to be revealed, or is simultaneous temporal processing an ability that may be revealed in standard conditions? The purpose of this section is to make the case that simultaneous temporal processing occurs in many conditioning procedures, including the most standard procedures such as delay and trace conditioning. Variations in the location of the reinforcer in a cycle, in stimulus durations, and cycle durations can also be understood as examples of simultaneous temporal processing.

Location of the reinforcer in a cycle

Two types of conditioning procedures studied in Pavlov's laboratory were delay conditioning and trace conditioning (Pavlov, 1927). In delay conditioning, a stimulus is presented for a fixed duration and a reinforcer is presented at the end; in trace conditioning, a stimulus is presented for a fixed duration and a reinforcer is presented at some fixed time after the termination of the stimulus.

The two procedures diagrammed at the top of Figure 5 are variants of the Pavlovian delay and trace conditioning procedures (unpublished research of M. MacInnis). Eighteen rats were trained in a box with one stimulus (white noise), one reinforcer (a food pellet), and one measured response (head entry into the food cup). A cycle consisted of 20 s with noise and 100 s without noise. These cycles continued throughout a session with food available at the same point during each cycle with a probability of 0.5. The data are shown for cycles in which food was delivered. Nine rats received the delay conditioning procedure before the trace procedure; nine other rats received the treatments in the other order. In the case of delay conditioning, food was available at the end of the stimulus on a random half of the cycles (indicated by the first arrow); in the case of trace conditioning, food was available 10 s after stimulus termination on a random half of the cycles (indicated by the second arrow).

The delay procedure provided three time markers (stimulus on, stimulus off, and food delivery). The time from stimulus onset to food availability was 20 s; the time from stimulus termination to food availability was 0 s; and the time from food to food was 120 s. In the delay procedure, the response rate increased as a function of time since stimulus onset; at stimulus termination the response rate declined abruptly; and there was a very small increase in the response rate as a function of time during the last 60 s of the cycle.

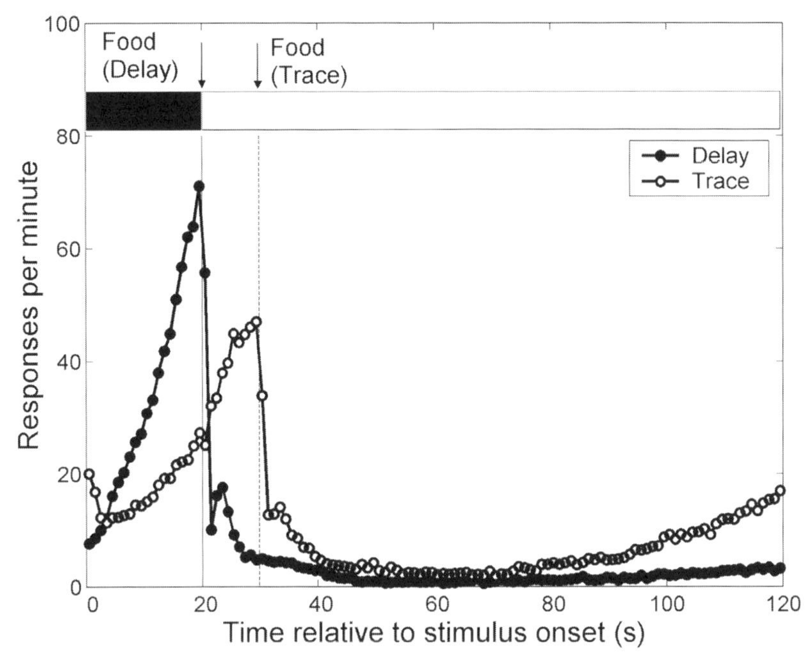

Figure 5: Response rate as a function of time from stimulus onset under a delay and trace conditioning procedure. The procedure is shown in the top of the figure. Based on unpublished research of M. MacInnis.

The trace procedure also provided three time markers (stimulus on, stimulus off, and food delivery). The time from stimulus onset to food availability was 30 s; the time from stimulus termination to food availability was 10 s, and the time from food to food was 120 s. Simultaneous timing during the cycle was apparent in the pattern of results shown in Figure 5. In the delay condition there was an abrupt increase in response rate at the onset of the stimulus, a sharp dropoff in response rate at stimulus termination and food delivery, which was followed by a slow rise toward the end of the cycle. In the trace condition there was an increase in response rate at the onset of the stimulus; at stimulus termination it continued to increase, but with a somewhat steeper slope; and at the time of food delivery the response rate declined abruptly, and there was an increase in the response rate as a function of time in the last 60 s of the cycle. The multiple slopes present in the response gradients suggest that the animals were using more than just the food-to-food interval to determine how fast to respond. In addition, the dip in response rate present at the beginning of the cycle for the trace procedure is similar to the dips seen in the segmented peak interval, and presumably is another example of simultaneous temporal processing.

Variations in the stimulus duration and cycle duration

Two variables that affect the speed of acquisition and asymptotic performance of conditioned responses are the duration of the stimulus and the duration of the cycle. The duration of the stimulus is sometimes referred to as the duration of "the trial;" and the duration of the cycle is referred to as the duration of "the trial" plus the duration of "the intertrial interval." Conditioning may be improved by reducing the duration of the stimulus or increasing the duration of the cycle. The speed of acquisition of autoshaping by pigeons has been found to be approximately the same when the stimulus to cycle duration ratio is the same (Gibbon, Baldock, Locurto, Gold, & Terrace, 1977). This is an example of timescale invariance in which relative, rather than absolute, time intervals control behavior (Gallistel & Gibbon, 2000, 2002).

The effect of variations in the stimulus and cycle duration can be studied with rats using the head entry response. Three recent studies have suggested that both stimulus and cycle durations are relevant, that the ratio of the two is a better predictor of performance than either one alone, but that timescale invariance is only an approximation (Lattal, 1999; Holland, 2000; Kirkpatrick & Church, 2000). These results provide evidence for simultaneous temporal processing.

The general procedure is illustrated at the top of Figure 6. For each animal, the interval between successive deliveries of foods is fixed (cycle duration), and the interval between stimulus onset and food is fixed (stimulus duration). Some results from the experiments of Kirkpatrick and Church (2000) are shown in Figure 6. The response rate during the stimulus as a function from stimulus onset is shown for groups with the same cycle duration of 180 s but different stimulus durations and also for groups with the same stimulus duration of 60 s but with different cycle durations. When the cycle duration was 180 s, the temporal gradients were ordered by the stimulus duration (15, 30, 60, and 120 s), and when the stimulus duration was 60 s, the temporal gradients were ordered by cycle duration (90, 180, and 360 s).

The contribution of simultaneous temporal processing is evident in the bottom two panels of the figure, which display the full gradients over the course of the cycle for pairs of groups that received stimulus/cycle duration ratios of .67 or .17. The gradients were not the same for two conditions in which the stimulus/cycle ratio was .67, and they were not the same for two conditions in which the stimulus/cycle ratio was .17. The time of stimulus onset is marked on each function by an arrow. Timing from the prior food delivery can be seen particularly well in the groups with the .17 ratios. Response rates increased gradually prior to stimulus onset and then abruptly changed at stimulus onset so that responding increased more rapidly.

This procedure gave the animal two time cues to use, in order to anticipate when the food would be made available: Stimulus on and food delivery. (Food was always delivered at the time of stimulus termination so this was not a differential cue.) The observed performance is apparently affected by the simultaneous temporal processing of these two intervals. The rats evidently timed from both cues, as seen in the bottom panels of Figure 6, and the response during the stimulus may have been determined by a combined influence of timing from both markers. When the cycle duration was constant, then the additional effect of the stimulus was determined directly by stimulus duration. However, when stimulus duration was held constant (stimulus 60 s groups), variations in the cycle duration might have resulted in differences in the additional amount of responding during the stimulus. A shorter cycle would have produced more responses from the prior food delivery and thus there would be a greater response rate during the stimulus if the rat was simultaneously timing both cues at once.

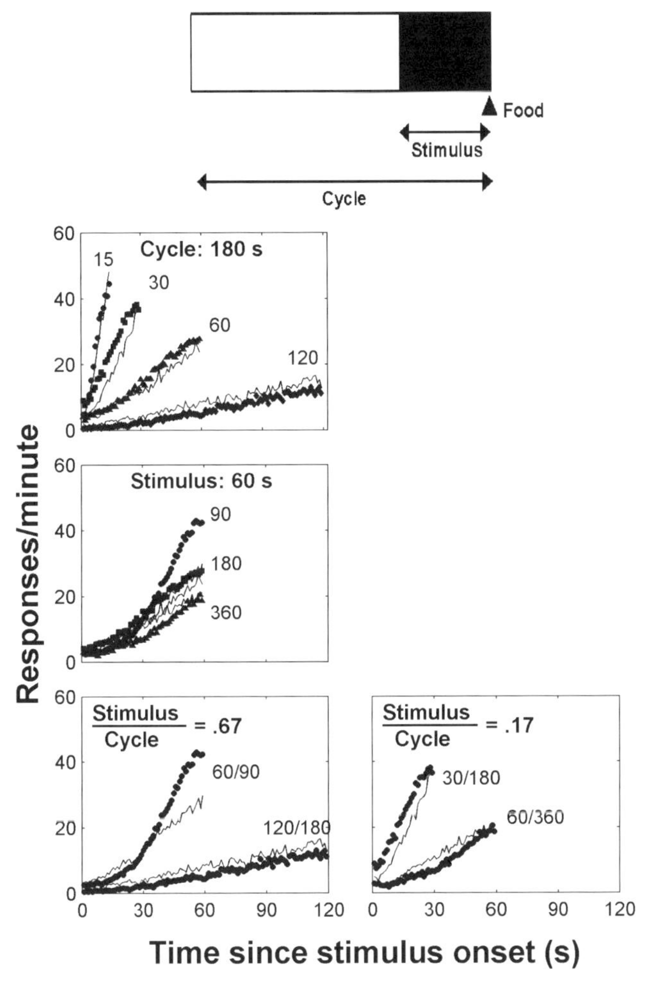

Figure 6: Effect of variations in duration of stimulus and cycle on the response rate as a function of time. Top panel is the procedure. Middle left panel: Effect of stimulus duration with a fixed cycle; Middle right panel: Effect of cycle duration for a fixed stimulus; Bottom data panels: Differences in response rate gradients with constant stimulus/cycle ratios. The thin smooth lines are based on packet theory; the arrows indicate times of stimulus onset (see text). Redrawn from Kirkpatrick and Church (2000).

A formal model of the process, packet theory, is being developed to account for the behavior of rats in procedures involving one or more temporal cues (Kirkpatrick, 2002; Kirkpatrick & Church, 2003). A basic idea is that behavior, such as head entry into a food cup, consists of bouts of responses. A distinction is made between a bout, which is an observed dependent variable, and a packet, which is an intervening variable of the theory. The proposal is that the momentary probability of producing a packet is controlled by the expected time to food. Figure 7 shows an overview of the theory in which the stimulus (stimulus to food interval) was 60 s and the cycle (food to food interval) was 180 s. The top left panel shows the expected time to food, E(t), decreasing from 60 s to 0 s as a function of the time since stimulus onset. The middle left panel shows the same function in memory. If the interval between stimulus and food was not always 60 s, a standard weighted linear combination rule was used to combine the most recent perceived function with the remembered function. The bottom left panel shows a decision function that is determined by the memory function. It is inverted in direction, and normalized to produce a unit area. The three panels on the right provide the same information for the food-to-food interval, and the bottom panel shows the two functions on the timeline for a cycle. As shown in the procedure (at the top of Figure 6) with a 60-s stimulus at the end of a 180-s cycle, the stimulus begins at 120 s. The dependent variable is the probability of a packet of responses. One plausible combination rule for the two functions is a simple summation, which was used.

The results of a simulation of the packet model (with the same parameter settings as used by Kirkpatrick, 2002) is shown by the thin lines near the data points in Figure 6. With these parameters, packet theory provides a good approximation of the effect of stimulus duration, and a good approximation of the failure of constancy of the stimulus/cycle ratio, but it does not show the degree to which the duration of the cycle affected the response rate during the stimulus. This may be improved by increasing the weighting of the cycle effect, or it may indicate that the combination rule is incorrect.

In this analysis, a quantitative theory of timing was used to predict the response rate as a function of time from time markers, and then a simple summation of the response rates was used as the combination rule. There are many plausible ways to combine information from two sources about the time to the next food. One could combine the times in various ways (sum, mean, larger, smaller, etc.) and these operations could be done on any transformation of the timescale. Alternatively, one could use each of the time estimates to the next food to generate response probabilities that could be assumed to be independent and combined in various ways to produce logical outputs such as "and" or "or." There may be a typical combination rule, or it may be that the combination rule is based on the task demands. Both an empirical and theoretical approach can be useful. The goal is to be able to predict what an animal will do with multiple sources of information about the time to reinforcement, based on knowledge of what the animal will do with each of the sources of information individually.

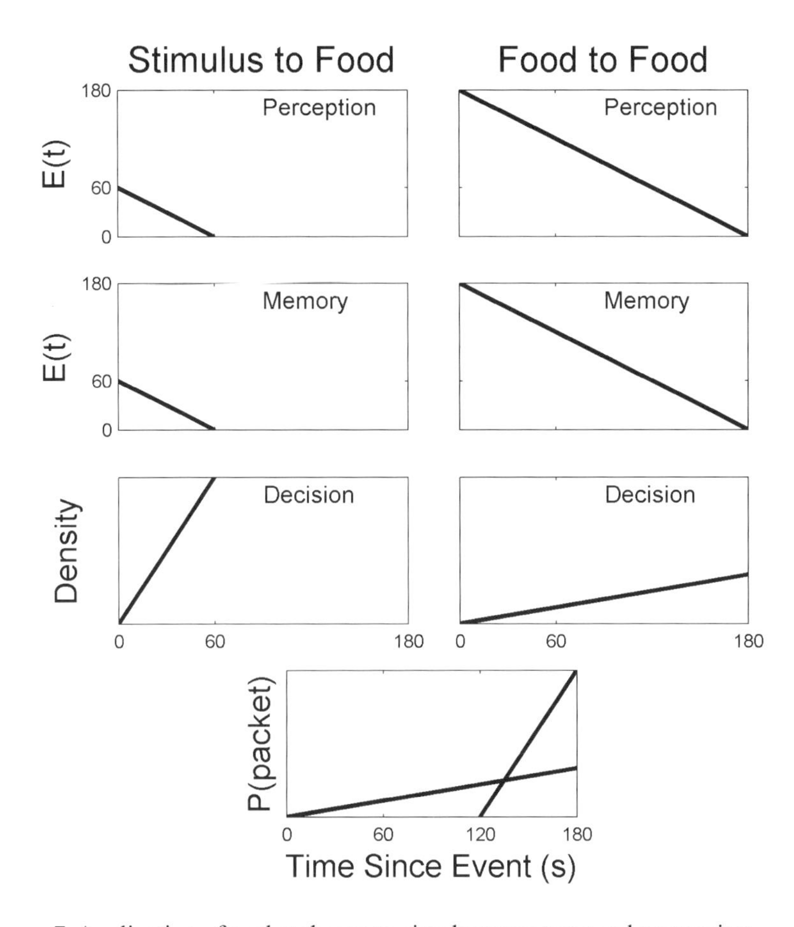

Figure 7: Application of packet theory to simultaneous temporal processing. Two independent functions are produced by stimulus onset and food onset. The two functions, shown in the bottom panel, produce the probability of a packet of responses according to an additive combination rule. This analysis is based on Kirkpatrick (2002).

Conclusions

The demonstration of simultaneous temporal processing does not require the use of any specialized procedures, such as the segmented fixed-interval procedure. It is likely present in many, if not most, of the procedures used in the study of conditioning and instrumental learning. Although the examples in this chapter were based on the use of constant intervals, introducing variability into the intervals does not eliminate the timing of them (Kirkpatrick & Church, 2003).

Much more work is needed in the development, evaluation, and selection of a theory of simultaneous temporal processing. At present there is no generally accepted quantitative theory of simultaneous temporal processing. Scalar timing theory and packet theory are two candidates, but (perhaps with some modifications) many other quantitative theories of conditioning and timing are also candidate theories that may be developed. An essential feature of any theory of simultaneous temporal processing is the selection of a combination rule for the multiple sources of temporal information.

The criteria for the evaluation of a theory includes much more than goodness of fit. A good theory should be relatively inflexible so that it cannot fit a large number of data patterns that do not occur. Quantitative measures that take into account the inflexibility of a theory are readily available to psychologists. Generality of a theory is also important. This sometimes refers to the formal characteristic of separation of fits of replicable factors from the fits of error factors; but it may also be used to refer to the more informal characteristic of fitting data from many sources. The growth of the availability of data archives (Kurtzman, Church, & Crystal, 2002) and secondary data analysis, may make it feasible to apply a theory of simultaneous temporal processing to a wide range of experimental data.

References

Catania, C. A., & Reynolds, G. S. (1968). A quantitative analysis of the responding maintained by interval schedules of reinforcement. *Journal of the Experimental Analysis of Behavior, 11,* 327–383.

Church, R. M. (1984). Properties of the internal clock. *Annals of the New York Academy of Sciences, 423,* 566–582.

Church, R. M. (2003). A concise introduction to scalar timing theory. In W. H. Meck (Ed.) *Functional and neural mechanisms of interval timing* (pp. 3–22). Boca Raton, FL: CRC Press.

Church, R. M., & Gibbon, J. (1982). Temporal generalization. *Journal of Experimental Psychology: Animal Behavior Processes, 8,* 165–186.

Dews, P. B. (1962). The effect of multiple S^{Δ} periods on responding on a fixed-interval schedule. *Journal of the Experimental Analysis of Behavior, 5,* 369–374.

Dews, P. B. (1965a). The effect of multiple S^{Δ} periods on responding on a fixed-interval schedule: II. In a primate. *Journal of the Experimental Analysis of Behavior, 8,* 53–54.

Dews, P. B. (1965b). The effect of multiple S^{Δ} periods on responding on a fixed-interval schedule. III. Effect of changes in pattern of interruptions, parameters and stimuli. *Journal of the Experimental Analysis of Behavior, 8,* 427–435.

Dews, P. B. (1966). The effect of multiple S^{Δ} periods on responding on a fixed-interval schedule. IV. Effect of continuous S^{Δ} with only short S^{D} probes. *Journal of the Experimental Analysis of Behavior, 9,* 147–151.

Dews, P. B. (1970). The theory of fixed-interval responding. In W. N. Schoenfeld (Ed) *The theory of reinforcement schedules* (pp. 43–61). New York: Appleton-Century-Crofts.

Gallistel, C. R., & Gibbon, J. (2000). Time, rate and conditioning. *Psychological Review, 107,* 289–344.

Gallistel, C. R., & Gibbon, J. (2002). *The symbolic foundations of conditioned behavior.* Mahwah, NJ: Erlbaum Associates.

Gibbon, J. (1977). Scalar expectancy theory and Weber's Law in animal timing. *Psychological Review, 84,* 279–325.

Gibbon, J., Baldock, M. D., Locurto, C. M., Gold, L., & Terrace, H. S. (1977). Trial and intertrial durations in autoshaping. *Journal of Experimental Psychology: Animal Behavior Processes, 3,* 264–284.

Gibbon, J., Church, R. M., & Meck, W. H. (1984). Scalar timing in memory. In J. Gibbon & L. G. Allan (Eds.), *Annals of the New York Academy of Sciences: Timing and time perception, 432* (pp. 52–77). New York: New York Academy of Sciences.

Holland, P. C. (2000). Trial and intertrial durations in appetitive conditioning in rats. *Animal Learning and Behavior, 28,* 121–135.

Kelleher, R. T. (1966). Conditioned reinforcement in second-order schedules. *Journal of the Experimental Analysis of Behavior, 9,* 475–485.

Keller, F. S., & Schoenfeld, W. N. (1950). Principles of psychology. New York: Appleton-Century–Crofts.

Kirkpatrick, K. (2002). Packet theory of conditioning and timing. *Behavioural Processes, 57,* 89–106.

Kirkpatrick, K., & Church, R. M. (2000). Independent effects of stimulus and cycle duration in conditioning: The role of timing processes. *Animal Learning and Behavior, 28,* 373–388.

Kirkpatrick, K., & Church, R. M. (2003). Tracking of the expected time to reinforcement in temporal conditioning procedures. *Learning and Behavior*, 31, 3–21.

Kurtzman, H. S., Church, R. M., & Crystal, J. D. (2002). Data achieving for animal cognition research: Report of an NIMH workshop. *Animal Learning and Behavior, 30,* 405–412.

Lattal, K. M. (1999). Trial and intertrial durations in Pavlovian conditioning: Issues of learning and performance. *Journal of Experimental Psychology: Animal Behavior Processes, 25,* 433–450.

Meck, W. H. (1987). Vasopressin metabolite neuropeptide facilitates simultaneous temporal processing. *Behavioural Brain Research, 23,* 147–157.

Meck, W. H., & Church, R. M. (1984). Simultaneous temporal processing. *Journal of Experimental Psychology: Animal Behavior Processes, 10,* 1–29.

Meck, W. H., Williams, C. L. (1997). Simultaneous temporal processing is sensitive to prenatal choline availability in mature and aged rats. *Neuroreport: An International Journal for the Rapid Communication of Research in Neuroscience, 8,* 3045–3051.

Olton, D. S., Wenk, G. L., Church, R. M., & Meck, W. H. (1989). Attention and the frontal cortex as examined by simultaneous temporal processing. *Neuropsychologia, 26,* 307–318.

Pavlov, I. P. (1927). *Conditioned reflexes.* London: Oxford University Press.

Stubbs, A. (1968). The discrimination of stimulus duration by pigeons. *Journal of the Experimental Analysis of Behavior, 11*, 223–228.

Stubbs, D. A. (1971). Second-order schedules and the problem of conditioned reinforcement. *Journal of the Experimental Analysis of Behavior, 16*, 289–313.

Chapter 2:
Applying the scalar timing model to human time psychology: Progress and challenges

JOHN H. WEARDEN

Abstract

Scalar timing (or scalar expectancy) theory, SET, was originally developed as an explanation of the performance of animals on temporally-constrained reinforcement schedules. In the last decade, however, it has had a growing influence on the study of human timing, and may now even be the dominant approach. The present paper reviews its successes, and points to some challenges for the future. Among the successes are (1) a re-invigoration of the old idea that some aspects of timing in humans depend on an "internal clock," (2) the provision of a framework for developmental studies of timing in humans (both of children and the elderly), (3) the development of precise quantitative models of timing in humans, which depend on an interaction of clock, memory, and decision processes. In spite of these successes, however, many problems remain. Some of these concern details of how the SET system itself works, particularly questions concerning the roles of memory and decision processes. Data from recent experiments which manipulate memory and decision mechanisms in the SET model, in an attempt to clarify their operation, will be presented. Another set of problems concerns the application of SET-related ideas to "classical" timing tasks such as production, reproduction, and verbal estimation. Behavior on this "classic trio" of tasks often seems at variance with the scalar model, but it will be shown that the incompatibility may be more apparent than real, and that SET-based models may be used to explain many aspects of performance on these classical tasks. Thus, not only can SET-based models account for recently-collected data on many timing tasks, but they may be able to provide the first rigorous models of behavior on procedures known for more than 150 years.

Introduction

The application of *scalar timing* theory (or *scalar expectancy* theory, SET) to human time psychology is one of the (rather few) success stories of the long history of attempts to apply models developed initially with non-human animals to the behavior of humans. SET was developed initially to explain the striking regularities in the behavior of rats and pigeons both on classical schedules of reinforcement with some sort of temporal constraint, and on specially designed timing tasks such as bisection (Church & DeLuty, 1977), and temporal generalization (Church & Gibbon, 1982).

An article by Wearden and McShane (1988) was probably the first to analyze human timing performance using ideas derived from SET, although the possibility of doing this had been mentioned before by others. Since then, perhaps as many as 100 journal articles and chapters have applied ideas derived from SET to human timing, and a recent review by Allan (1998) discusses many of them. Lack of space precludes a complete exposition of SET here, but non-technical discussions of the operation of SET are available in Wearden (1994, 1999, 2001).

After this initial introduction, the remainder of this chapter is organized into three main sections. The first (*Progress*) discusses what seem to me to be examples of unalloyed progress in understanding human timing made possible by SET. The second (*Progress and Challenges*) discusses some areas where, although SET has made major contributions to increasing understanding, important problems remain unsolved. The final section (*Challenges*) discusses areas which appear problematical for SET, and focuses particularly on one of them, the application of SET to "classical" timing tasks.

Although some aspects of the detailed mechanics of SET will be mentioned at various points in this chapter, two features of the SET model need to be mentioned in this preface. Firstly, when we say that some behavior conforms or does not conform to SET, what do we mean by this? SET requires that time representations have two properties. The first of these is *mean accuracy*: the requirement that internal "estimates" of some real time duration, t, are on average accurate. The second is the *scalar property* which gives the theory its name. This is essentially a form of Weber's law, which can be exemplified in various ways. Perhaps the simplest of these is the requirement that the standard deviation of judgments of time grows as a linear function of the mean, or alternatively that the coefficient of variation statistic (standard deviation/mean) remains constant as the interval to be timed varies. Another way of testing the scalar property is to inspect the data for the property of *superimposition* (called *superposition* in U.S. English), the requirement that measures of timing obtained when different absolute durations are timed superimpose when plotted on the same relative scale. Thus, behavior which does not conform to mean accuracy or the scalar property seems, at least at first sight, incompatible with SET.

The other feature of SET which needs to be mentioned here is the structure of its well-known information-processing variant. This embodies the SET system in a tripartite structure. The first component is a pacemaker-accumulator internal clock which provides "raw" representations of durations. The second is a memory system consisting of a short-term working memory reflecting accumulator contents, and the second is a reference memory component containing some kind of representation of "important" times such as those associated with reinforcement in experiments with animals or used as standards in experiments with humans. Finally, a decision component is needed to produce observed behavior, and this usually involves comparisons of the contents of working memory with standards stored in reference memory. As will be seen below, some progress has been made isolating these different parts of the SET system, although many things currently remain obscure.

Progress

I would like to concentrate on two aspects of the application of the SET system to human timing that I personally consider to have been particularly fruitful. The first is the fact that SET has provided coherent quantitative models of behavior on some timing tasks that have proved useful in understanding the ways in which different groups (e.g., children, student-age adults, and the elderly) differs. The second is the fact that SET has revitalized the old idea that humans might perform on some timing tasks by using an internal clock.

Pre-SET time psychology is filled (one might almost say "littered") with studies comparing group X and group Y on some timing task and finding that they behave differently. So, children and young adults differ, people with brain damage differ from their intact controls, things happen to timing performance as people get older.....but so what? Merely demonstrating that two subject groups differ in timing performance is almost spectacularly uninformative, as it gives us no insight whatsoever into what mechanisms might underlie the performance differences observed. By providing quantitative models of performance on (some) timing tasks, SET lets us go beneath the skin of the subject and at least make reasonable conjectures as to what mechanisms might be involved in performance differences.

Perhaps the simplest examples come from the study of temporal generalization performance in different groups. Temporal generalization, developed originally as a method for studying timing in rats by Church and Gibbon (1982), has been extensively applied to humans, usually in variants of a method developed by Wearden (1991a, 1992). Humans are initially presented with some duration identified as a standard (e.g., a tone 400 ms long). They then receive other durations, shorter or longer than the standard, or equal to it, and simply have to judge whether or not each duration was the standard by responding YES or NO. The proportion of YES responses plotted against stimulus duration is a temporal generalization gradient. The conventional analysis of human behavior on this task is the "modified Church and Gibbon," MCG, model, developed by Wearden (1992) from Church and Gibbon (1982).

On temporal generalization, the standard duration (s) is assumed to be stored in reference memory, and represented as a Gaussian distribution with a mean s and some coefficient of variation c, essentially representing the precision with which the standard duration is stored. Each comparison duration, t, is assumed to be timed without error, and the MCG model produces a YES response when $abs(t - s^*)/t < b^*$, where s^* is a sample drawn from the reference memory (which differs on each trial), and b^* is a sample drawn from a threshold distribution with mean b, and some standard deviation which is usually $0.5b$ (*abs* indicates absolute value).

If different groups differ in performance on temporal generalization performance then, according to the MCG model, they might differ in c, the "fuzziness" of the representation of the standard, or in b, the decision threshold. Wearden, Wearden, and Rabbitt (1997a) tested the performance of elderly subjects (60–80 years) on temporal generalization, and looked at effects of both age and IQ. More recently, Droit-Volet, Clément, and Wearden (2001) examined the temporal generalization performance of children of 3, 5, and 8 years, and several studies (Wearden, 1992; Wearden, Denovan, Fakhri, & Haworth, 1997b) have obtained data from student-age subjects. In general, "precision" of performance in terms of the width of the temporal generalization gradient was greatest in students, and poorer both in children and in the elderly. The MCG model suggested that the principal cause was the coefficient of variation, c, of the memory representation of the standard, and Figure 1 shows c plotted against subject age, with student data averaged over a number of conditions and assuming a subject age of 21 years.

The effects are obvious: children of 3 and 5 had the highest c values, these dipped markedly at the age of 8, and continued to dip towards student age, only to slowly rise again in the groups averaging 65 and 75 years. To be strictly accurate, some other developmental trends were present as well, such as a tendency for the youngest children to respond YES or NO randomly (Droit-Volet et al., 2001), but the essential point here is how the quantitative model derived from SET enables theoretically meaningful comparisons of the behavior of different groups, thus going much further than the usual final conclusion that different groups differ in performance.....but we can't say any more. The old adage that "nothing is so practical as a good theory" seems highly applicable here, and if SET-compatible tasks were used with, say, patients with brain damage then we might be able to draw conclusions stronger than those presently available.

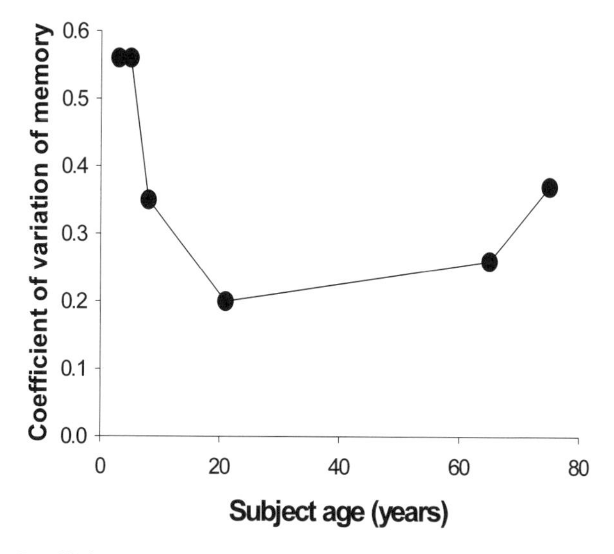

Figure 1: Coefficients of variation of memory representation of standard durations obtained from fits of the MCG model to temporal generalization performance, plotted against subject age. Data come from Droit-Volet, Clément, & Wearden (2001), Wearden (1992), and Wearden et al. (1997a, b).

The second area of progress for SET has been a revitalization of the old idea, traceable at least back to François (1927) that humans possess some sort of *internal clock*. Once again, space constraints mean that I cannot do justice to this important notion, which has spawned much diverse research. Wearden and Penton-Voak (1995) reviewed the bizarre field of body temperature and time estimation, and found general support for internal clock ideas. Luckily, Treisman, Faulker, Naish, and Brogan (1990) developed, as a by product of their work which was actually focused on something slightly different, an innocuous method of apparently changing the pacemaker rate of an internal clock in humans, the use of repetitive stimulation in the form of clicks or flashes.

Penton-Voak, Edwards, Percival, and Wearden (1996) used a variant of Treisman et al.'s method, and showed that people estimated the duration of both auditory and visual stimuli as longer when these were preceded by clicks than when presented alone. Furthermore, the click effect was multiplicative, that is, greater at longer durations than shorter ones, exactly the effect predicted by an increase in pacemaker speed. Penton-Voak et al. also demonstrated that, while the subjective duration of stimuli estimated was lengthened by clicks, intervals produced by the subjects themselves were shortened. This again is consistent with the idea that humans possess an internal clock: If some duration produced is normally associated with x ticks of the internal clock, then when the pacemaker is speeded up the x ticks required will be reached in a shorter real time, thus shortening intervals produced. Burle and Bonnet (1999) and Burle and Casini (2001) have also demonstrated this shortening of intervals produced by click trains.

A recent article by Droit-Volet and Wearden (2002) used repetitive flicker to apparently speed up the internal clock of children as young as 3 years of age. The experimental details are too complicated to be described here, but once again a multiplicative effect of flicker, consistent with a change in pacemaker speed, was obtained. As Droit-Volet and Wearden (2002) note, obtaining such an effect in young children implies that the effect is a very "primitive" one, operating on some fundamental, perhaps even biological, level, rather than affecting complex response strategies.

Internal clock models consistent with the pacemaker-accumulator model of SET have also recently been used (Wearden, Edwards, Fakhri, & Percival, 1998) to account for the well-known modality effect in timing, that "tones are judged longer than lights": In other words, auditory stimuli appear to last longer than visual ones of the same real duration. Once again, the modality effect was greater at longer durations than shorter ones, indicating a pacemaker speed difference between the conditions.

It may be that one day the physiological mechanism of human timing will be uncovered (although my personal view is that this discovery is not imminent), and the idea of a pacemaker-accumulator internal clock shown to be invalid. However, one might say that if people don't have a pacemaker-accumulator internal clock, they certainly *behave* as if they do, so any model which replaces this clock idea will have to account for the data which seems to support pacemaker-accumulator clocks so compellingly.

Progress and challenges

In the second section of this article, I will discuss some areas where ideas compatible with SET have stimulated research, but in which serious outstanding problems remain. These relate to the tripartite clock-memory-decision structure of SET, and some critics (e.g., Staddon & Higa, 1999) have complained that this structure allows SET too much flexibility in fitting data, rendering it almost undisprovable.

An important aspect of SET is that it has stressed how behavior on all timing tasks results from more than just properties of an internal clock. Even if some clock produces "raw" representations of time, these must be temporarily stored and compared with various sorts of standards by using some decision process: Behavior does not necessarily directly reflect the properties of underlying representations of time. But, as I pointed out elsewhere (Wearden, 1999), it is difficult to know when the fit obtained by varying the memory and decision mechanisms of SET so that the predictions of the model converge on data is a reasonable one, as different memory and decision process could always be posited.

To partly address this charge of undisprovability, Wearden (1999) suggested the program of systematically trying to "isolate" the different components of the SET system, to see if they behaved as standard versions of SET require. Part of the program of isolating the internal clock component had already been carried out with animals (e.g., Meck, 1983), and the section above showed how definite evidence for the sort of pacemaker-accumulator clock proposed by SET had been obtained from studies with humans. The other parts of the SET system have received far less attention.

The memory level of SET consists of two memory stores, one is a short-term *working memory*, assumed to represent the contents of the accumulator more or less accurately. The second is a *reference memory*, which stores "important" times necessary for the task in hand.

Considering the working memory first, it is currently unclear what relation the working memory of SET bears to other accounts of working memory, such as those proposed influentially by Baddeley (1986). A second problem is that there is some evidence that short-term memory of the duration of events is susceptible to an unusual form of forgetting, "subjective shortening," a tendency for the duration represented to become subjectively shorter (rather than to degrade randomly) as the memory ages. Spetch and Wilkie (1983) popularized this effect using data from pigeons, but Wearden and Ferrara (1993) and Wearden, Parry, and Stamp (2002) have demonstrated it in humans and in unpublished studies with elderly people, children, and patients with Parkinson's Disease, I have replicated their effects with non-student subject populations. However, in spite of the apparent robustness of subjective shortening in humans it is currently unclear (1) why subjective shortening occurs at all, (2) how it relates to "conventional" short-term memory (see discussion in Wearden et al., 2002), and (3) how it might affect performance on the sort of tasks used to evaluate SET.

The reference memory component of SET plays a more important role than working memory in most versions of SET. This is mainly because Gibbon, Church and Meck (1984) proposed that the reference memory is the source of the scalar property (conformity to Weber's law, constant coefficient of variation, etc.) so often observed in timed behavior. So, for example, data from temporal generalization in rats superimposes when plotted on the same relative scale when the standard duration is varied over values of 2, 4, and 8 s because the reference memory representations of 2, 4, and 8 s have scalar variance (Church & Gibbon, 1982). This contrasts with the commonsense idea that the raw time representations produced by the internal clock itself might have scalar variance, possibly causes by variations in pacemaker speed from trial to trial. Lack of space prevents a full discussion of the issues here (although Jones & Wearden, in press-a, provides one), but some recent studies with humans (Allan & Gerhardt, 2001; Wearden & Bray, 2001) have provided simple demonstrations of the existence of the scalar property of behavior in situations where it is unlikely that reference memory was ever used by the subjects (particularly Wearden & Bray, 2001, Experiment 3, where the use of reference memory seems impossible). This does not, of course, prove that reference memory is not a source of scalar variance when it is used, only that scalar variance can apparently arise from other sources.

The issue of what is actually stored in reference memory is also problematical. Timing experiments with animals often involve extensive training with "important" times, e.g., those associated with reinforcement, and it was natural for early work to assume that the repeated presentations of these important times were individually stored, each being slightly distorted, to form an extensive "memory distribution" in reference memory. In tasks with adult humans, training lasts seconds or minutes at most, so it seems unlikely that such a memory distribution could be formed. In addition, there is evidence from experiments with animals that once trained on some critical duration, t_1, animals can shift extremely rapidly to some new duration, t_2, suggesting that the "reference" memory can be rapidly overwritten (Higa, 1996; Lejeune, Ferrara, Simons, & Wearden, 1997).

Jones and Wearden (in press-b) attempted to manipulate the number of distinct "items" in reference memory in humans, using methods too complicated to be described here. Their conclusion was that temporal reference memory in humans, and possibly also in animals, did not consist of any extensive distribution of stored time values, but possibly only of a single "standard" which had upper and lower limits. So, for example, when resenting a person with instances of a 400-ms "standard" resulted in the following processes. The first 400 ms value, s, was multiplied by a constant with a mean of 1.0 and some variability, and the resulting value s^* stored. A subsequent presentation of the 400-ms standard would generate another potential representation of the standard, call this s'. If s' was within some percentage of the original s^*, say within 10% above and below, then essentially nothing happened, but if s' was outside this range, the reference memory was "perturbed" as s' was substituted for s^*. This "perturbation model" not only fitted data from humans well, but also accounted for many effects in experiments with animals.

Whether or not the perturbation model is correct, it serves to focus interest on the important issue of what temporal reference memory contains and how it changes when the duration to be timed changes, but remains constant when the duration remains constant.

Many other unresolved issues centre around the role of decision processes in the SET model. As Wearden (1999) pointed out, the decision processes specified for a particular timing task are usually conjectured ad hoc, albeit in a very plausible way, and will necessarily vary from one task to another. One problem is that many different models can fit the same data more or less equally well. For example, the MCG model outlined above works well for temporal generalization gradients in humans, but even the original paper proposing it (Wearden, 1992) showed that models assuming that the threshold had no variance, or that variance was presented in the just-presented duration, *t*, as well as the representation of the standard, worked just as well. Of course, these models are small variants of the original MCG model, but sometimes models which differ more radically (and which have quite different psychological implications) are hard to distinguish.

Bisection also furnishes a number of examples. On a bisection task, people initially receive examples of standard *short* and *long* durations (*S* and *L*, e.g., tones 200 and 800 ms long). They then have to make judgments about other durations, *S* and *L*, and values in between, usually in terms of their similarity to *S* or *L* (e.g., "was that duration more similar to the standard *short* or the standard *long* duration"). SET-compatible models developed by Allan and Gibbon (1991) and Wearden (1991b) both fitted the data well, but employed different rules for judging the similarity of some duration *t* to *S* and *L*. However, worse was to come, as Wearden and Ferrara (1995) suggested that people were not actually making any kind of comparison between *t* and *S* and *L*, but instead comparing each *t* with the *mean* of all the durations experienced, including *S* and *L*, but awarding no special status to these "standards." Wearden, Bajic, and Brocki (submitted for publication) have recently shown that a model incorporating a variant of this idea fits the overall body of data on temporal bisection in humans (derived in their article from 77 data sets) better than any published competitor. So, what is actually happening in temporal bisection, what is being used as a "reference" for timing decisions made, and what sort of decisions are operative? These questions still have no definite answers.

In psychology, we are so accustomed to theories accounting for data poorly that we are embarrassed when a theoretical system like SET generates not one model that fits data excellently, but any number of them! How can we decide between the many possibilities that changing the memory and decision processes of SET can throw up? Perhaps the only way is to look at variants of experimental procedures, and to see if the models proposed predict behavior correctly when some aspect of the experimental situation is changed. For example, Wearden and Ferrara (1995) showed that changing the spacing of intermediate durations in bisection (from linear to logarithmic spacing, for example) between *S* and *L* changed bisection performance (as did Allan, 2002). Although virtually any model can fit any single set of bisection data, it is not the case that all models can simulate such stimulus spacing effects. In fact, any model which simply assumes that the decision about some time value, *t*, depends simply on a comparison between *t* and *S* and *t* and *L* cannot in principle simulate such effects without adding additional factors to the model.

In this vein, Wearden and Grindrod (2003) tried to manipulate the decision processes made in temporal generalization by encouraging or discouraging people to make YES responses by the award and deduction of points. In their study, durations remained identical between different conditions, and the manipulation was intended to change decision processes alone. Unsurprisingly, the manipulation worked: People made more YES responses overall (particularly to durations close to the standard duration) when the contingencies encouraged this, but the central question is how the SET-based model which fits temporal generalization in humans, the MCG model discussed earlier, modelled data. If the standard interpretation according to SET is correct, Wearden and Grindrod's manipulation should change the response threshold, leaving everything else unchanged, and this prediction was supported reasonably well, in that the main effect was a change in the threshold value, rather than the coefficient of variation of memory representation of the standard.

In spite of the success of this simple manipulation, much more work needs to be done on variants (both small and large) of standard SET-compatible tasks like temporal generalization and bisection before we can be completely sure as to what people are doing when they perform on these tasks. Just because a simple model fits the (equally simple) data well does not mean that the model fitted is correct.

Challenges

There are some areas in which SET remains either untested, or highly problematical, even within the domain of the laboratory-based timing of short durations. I will mention three, but concentrate only on the last one, a personal favorite. The first challenge is to link SET with more conventional cognitive psychology, in particular that part which deals with memory and attention. Numerous studies have shown that attention plays a critical role in timing behavior (e.g., Brown & West, 1990; Macar, Grondin, & Casini, 1994), but it is unclear how what is basically a mechanical system like SET can accommodate the notoriously tricky concept of attention. SET, reflecting its roots in behaviorist psychology, regards animals, and humans, as processing systems making timing decisions according to the operation of a specified internal clockwork. The subject is just a vessel responding to stimuli and processing them to produce output. This characterization of SET, while perhaps bleak, is not intended to be in the least pejorative, as personally I completely agree with this conception, but how the ghostly concept of "attention" can be defined so as to be incorporated into quantitative models remains hard to imagine.

Another major challenge is to try to understand the physiological processes underlying timing. Recent articles (e.g., Mattel & Meck, 2000) have criticized the idea of a pacemaker-accumulator clock on physiological grounds. As I note above, timing behavior often conforms well to the mathematical predictions of a pacemaker-accumulator clock, so if it is physiologically implausible its successor must be able to account to data which seem so consistent with it. Oscillator-based models (Church & Broadbent, 1990; Mattel & Meck, 2000; Wearden & Doherty, 1995) are said to have the cachet of greater physiological plausibility than the pacemaker-accumulator clock, but whether then can predict behavior as well as the much-maligned clock remains to be seen.

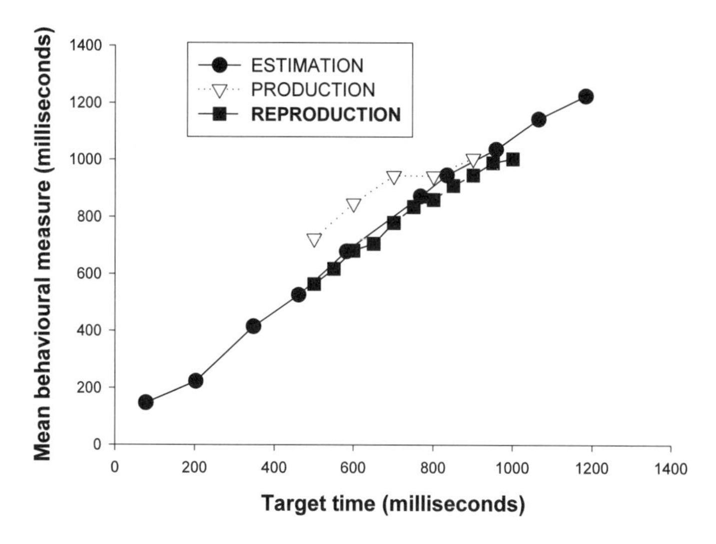

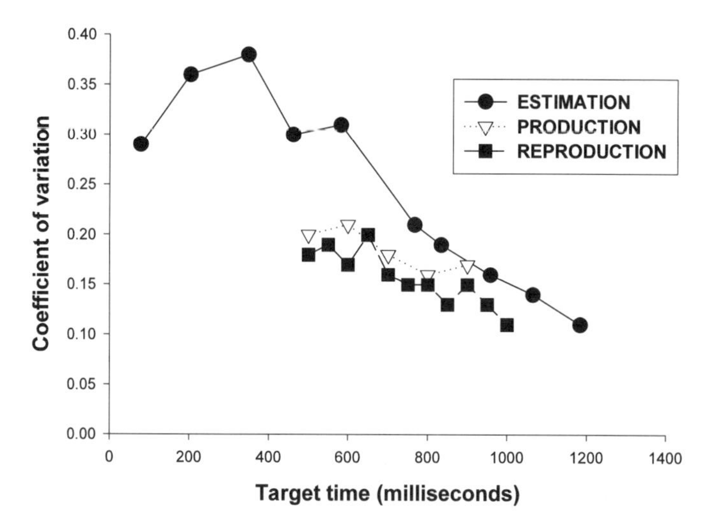

Figure 2: Upper panel: Mean measures of behavior plotted against target duration from 80–83 subjects performing on verbal estimation of tone durations, production of intervals without feedback, or temporal reproduction. Lower panel: coefficient of variation standard deviation/mean) of measures of behavior from verbal estimation, production, and reproduction.

My choice of a problem for the remainder of this section is neither of the two previously-mentioned, but rather that of the application of SET to "classical" timing procedures, in particular the production, reproduction, and verbal estimation of duration. It is perhaps ironic that while simple SET-compatible models can deal with what appear to be quite complicated experimental procedures (such as categorical timing, which involves making judgments about 18 or 24 different durations, see Wearden, 1995), we have few or no properly-developed models of how people produce, reproduce, or assign verbal labels to durations, in spite of the fact that such procedures have been used since the earliest days of time psychology (Fraisse, 1964). In fact, data from such procedures seem at first sight clearly incompatible with SET. Figures 2 and 3 show data from an experiment where 80–83 female undergraduate students (the number depending on the procedure used) produced short durations without feedback, reproduced durations, or assigned verbal labels to them.

The upper panel of Figure 2 shows the mean measures of behavior plotted against target time. Close to linear relations are evident in all the three cases, but the lower panel of Figure 2 shows that the scalar property of constant coefficient of variation (standard deviation/mean) is violated, sometimes quite severely, in data from all three procedures. Figure 3 shows the behavior measures expressed as a fraction of the target, with the lower panel showing the same data as the upper one but on a more sensitive scale. Obviously, measures of timed behavior from all three procedures exceeded target times at short target times, but became closer to targets at longer times. This is a form of Vierordt's law, the principle that short durations are "overestimated" whereas longer durations are "underestimated" with some indifference point between these two values. The data in Figure 3 never show "underestimation," although some other data from reproduction, to be discussed later, do.

Behavior on all three classical timing tasks thus appears to conform to neither of the properties of behavior required by SET by violating mean accuracy (and instead showing Vierordt-like effects), and also violating the scalar property of variance by showing decreasing coefficients of variation as the intervals timed get longer. Given that mean accuracy and the scalar property are very commonly found in human timing (e.g., Wearden, 1992, 1995; Wearden et al., 1997b; Wearden, Bajic, & Brocki, submitted) this state of affairs is surprising, even shocking. Can SET rise to the challenge of modelling data which seem at first sight to violate it in important ways?

In order to model data from the "classic trio" of timing tasks, some very difficult problems need to be solved, at least in part. One of these, relevant both to production and verbal estimation, is the problem of "scaling," that of how conventional time units such as seconds or milliseconds are related to subjective time, in both directions (time units to behavior in the case of production, subjective duration to time units in the case of verbal estimation). Another problem relevant to verbal estimation is that of "quantization," the fact that when people make judgments (e.g., in ms) of the duration of short stimuli, they do not use all possible values (in fact, around 90% of the estimates end in "00," and almost all the rest in "50"). Although we know that quantization occurs, we have no detailed models of how it is done, or how quantization affects time estimates produced. For example, one attractive possibility is that "raw" representations of stimulus durations exhibit both mean accuracy and the scalar property, but some quantization effect intervenes to produce the non-scalar character of estimates shown in verbal estimation data. This possibility is easy to propose, but producing an accurate model of verbal estimation may be difficult, as many different means of quantizing verbal estimates can be imagined and simulated.

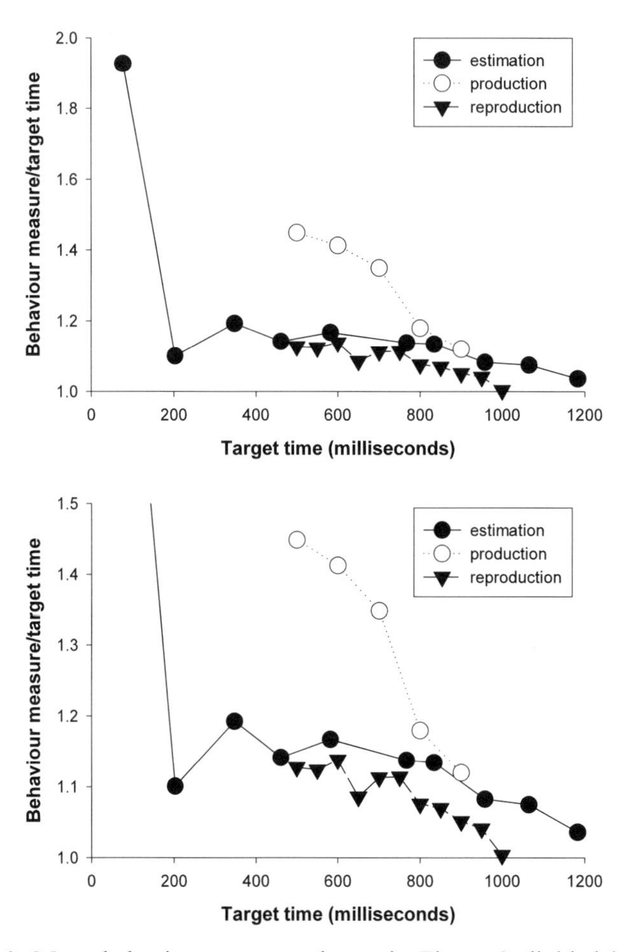

Figure 3: Mean behavior measures shown in Figure 2 divided by target time. Lower panel shows same data as upper one on a more sensitive scale.

The task of temporal reproduction offers a starting point for trying to develop SET-compatible models of performance that seems to violate SET, as people do not use conventional time units in reproduction (so the scaling problem is [probably] avoided), and the response, being motor, is not quantized in the same way as verbal estimates are. For the rest of this chapter, I will sketch out a potential model of temporal reproduction that reconciles performance with principles of SET.

First, some definitions. On temporal reproduction we will suppose that the subject initially receives some sample duration, s, in the form of a presented stimulus. The task is then to make some motor response, what I will call the reproduction of s, r, so that s and r are equal so far as the subject is concerned. The method used in my laboratory is what I call "reproduction by waiting" and involves just a single motor response. So, for example, s is defined by the time between two brief clicks, then a short random gap follows, then a third click. The subject's task is to wait after the third click for a time equal to s, the sample.

Figure 4 shows data from two independent replications of reproduction of sample durations ranging from 550 to 1050 ms. The upper panel of Figure 4 shows the mean times reproduced, which clearly vary linearly with sample duration, the centre panel shows that (unlike the reproduction data presented earlier) the coefficient of variation remained more or less constant, or declined only slightly, with increases in sample duration, and the bottom panel shows a Vierordt-type plot, where mean time reproduced is divided by the sample time.

Data from reproduction (Figures 2, 3 and 4) suggest in fact that r and s will not generally be the same, and that $r > s$ when s is small, and closer to it when s is longer (essentially Vierordt's law). To model reproduction, I will follow ideas used to explain repetitive tapping in the well-known model of Wing and Kristofferson (1973). The essential point is that the measured reproduction, r, is actually generated by two consecutive processes. The first process involves waiting until the elapsed time (from the third click) is "close enough" to s, at which time a response is initiated. Making the response, which takes some time, d, is the second process. So, the total reproduction, r, is made up of the time of response initiation (t) plus d. Given that the subjects in my experiments did not receive feedback, they had no way of knowing what the relation between r and s actually was.

There are various ways of implementing the ideas sketched above. A particularly simple version was incorporated into a computer model, which simulated the reproduction of target durations ranging from 500 to 1000 ms. On each trial, the sample duration, s, was multiplied by a random value drawn from a Gaussian distribution with mean 1.0 and coefficient of variation, c, to produce an effective standard for the trials, s^*. s^* varied from trial to trial, but was on average accurate (mean accuracy) and the multiplicative transform used also gave the representations of the sample the scalar property of variance. Thus, the basic representation of the time to be reproduced was completely compatible with SET principles. To initiate responding, the model ran the internal clock from zero until some time, t, which was "close enough" to s^*, in fact until the difference between s^* and t was some threshold proportion (in the simulations between 10% and 40%) of s^*. At this moment, the response was initiated, and the response was assumed to take on average 250 ms to be emitted, with the response time being represented as a value drawn from a uniform distribution running between 150 and 350 ms, the time on each trial being d. The total reproduction on the trial, r, was just $t + d$, whatever values they took on the trial.

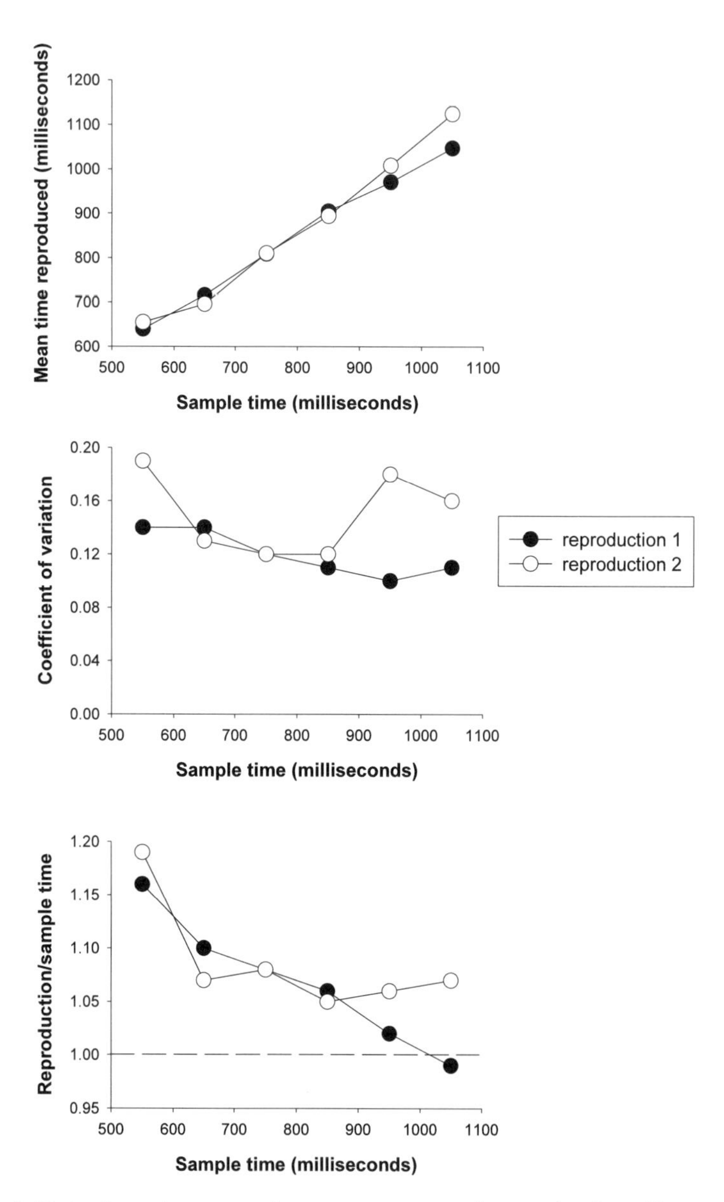

Figure 4: Data from two experiments on temporal reproduction. Upper panel: mean time reproduced plotted against sample (target) time. Centre panel: coefficient of variation of reproduced time plotted against sample time. Lowest panel: Mean time reproduced divided by sample time, plotted against sample time.

Figure 5 shows some simulations using this model, with 1000 trials at each time to be reproduced being simulated. *c*, the coefficient of variation of the memory of the sample duration, was kept at 0.2, and Figure 5 shows the effect of varying the response threshold. The upper panel shows the mean time reproduced as a function of the sample duration. It is clear that in all cases the model produced mean reproductions which grew linearly with sample duration. The lower panel shows a Vierordt-type plot, where the mean reproduction was divided by the target time. Here, values greater than 1.0 show what is (rather misleadingly) referred to as "overestimation" of the target, values less then 1.0 show "underestimation." Coefficients of variation remained roughly constant with changes in sample duration (thus resembling the data in Figure 4 but not those in Figure 2), but are not shown, to save space.

Two points are obvious from inspection of the lower panel of Figure 5. Firstly, Vierordt-type effects are found whatever the threshold value: mean times reproduced overshoot target times more when the targets are short than when they are long. Secondly, with some threshold values, short targets are "overestimated" while long targets are "underestimated" (e.g., threshold = 30 and 40%). Thirdly, with some threshold values there is an "indifference" point where a mean reproduction would be perfectly accurate: for the 30 and 40% thresholds this indifference point lies between 600 and 800 ms. The existence of an indifference point excited the interest of classical time psychologists to an excessive degree (e.g., Fraisse, 1964), and much psychological significance was attributed to it, but the present simulations suggest that the indifference point could be anywhere with choice of an appropriate threshold, and need not imply and profound difference between durations which are "overestimated" and those which are "underestimated."

It should be acknowledged that this simple reproduction model is not the only one that could be invented, although it is perhaps one of the simplest. In addition, it deals better with situations in which the coefficient of variation of times reproduced remains roughly constant. Marked decreases in coefficient of variation with increases in sample duration pose problems for the model, which can only simulate them by having very high (perhaps unrealistically high) variance in *d* from trial to trial. Nevertheless, the model performs the useful function of showing that Vierordt-type deviations from mean accuracy need not imply underlying non-scalar timing processes, but also begins the task of reconciling SET with classical timing data, and opening up a route to explaining why data collected since the earliest days of psychology have the form they do.

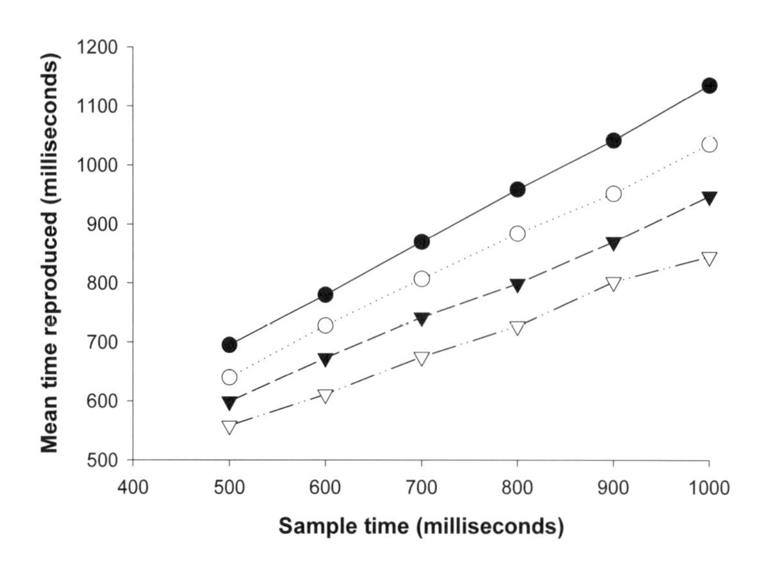

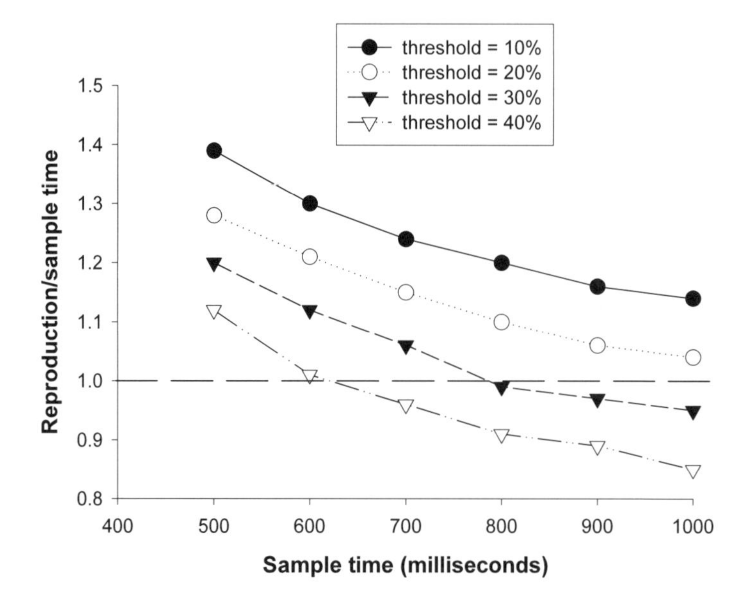

Figure 5: Upper panel: Mean time reproduced plotted against sample time derived from simulation discussed in text. Lower panel: Mean time reproduced divided by sample time, plotted against sample time, derived from simulation discussed in text.

References

Allan, L. G. (1998). The influence of the scalar timing model on human timing research. *Behavioural Processes*, *44*, 101–117.
Allan, L. G. (2002). The location and interpretation of the bisection point. *Quarterly Journal of Experimental Psychology, 55B*, 43–60.
Allan, L. G., & Gerhardt, K. (2001). Temporal bisection with trial referents. *Perception and Psychophysics, 63*, 524–540.
Allan, L. G., & Gibbon, J. (1991). Human bisection at the geometric mean. *Learning and Motivation*, *22*, 39–58.
Baddeley, A. D. (1986). *Working memory*. Oxford: Oxford University Press.
Brown, S. W., & West, A. N. (1990). Multiple timing and the allocation of attention. *Acta Psychologia*, *75*, 103–121.
Burle, B., & Bonnet, M. (1999). What's an internal clock for? From temporal information processing to the temporal processing of information. *Behavioural Processes*, *45*, 59–72.
Burle, B., & Casini, L. (2001). Dissociation between activation and attention effects in time estimation: Implications for clock models. *Journal of Experimental Psychology: Human Perception and Performance*, *27*, 195–205.
Church, R. M., & Broadbent, H. (1990). Alternative representations of time, number, and rate. *Cognition*, *37*, 55–81.
Church, R. M., & Deluty, M. Z. (1977). Bisection of temporal intervals. *Journal of Experimental Psychology: Animal Behavior Processes*, *3*, 216–228.
Church, R. M., & Gibbon, J. (1982). Temporal generalization. *Journal of Experimental Psychology: Animal Behavior Processes*, *8*, 165–186.
Droit-Volet, S., & Wearden, J. (2002). Speeding up an internal clock in children? Effects of visual flicker on subjective duration. *Quarterly Journal of Experimental Psychology, 55B,* 193–211.
Droit-Volet, S., Clément, A., & Wearden, J. H. (2001). Temporal generalization in 3- to 8-year-old children. *Journal of Experimental Child Psychology, 80,* 271–288.
Fraisse, P. (1964). *The psychology of time*. London: Eyre and Spottiswoode.
François, M. (1927). Contributions à l'étude du sens du temps: La température interne comme facteur de variation de l'appréciation subjective des durées. *L'Année Psychologique, 27*, 186–204.
Gibbon, J., Church, R. M., & Meck, W. (1984). Scalar timing in memory. In J. Gibbon and L. Allan (Eds.), *Annals of the New York Academy of Sciences, 423: Timing and time perception* (pp. 52–77). New York: New York Academy of Sciences.
Higa, J. J. (1996). Dynamics of time discrimination: II. The effects of mutliple impulses. *Journal of the Experimental Analysis of Behavior*, *66*, 117–134.
Jones, L. A., & Wearden, J. H. (in press-a). Double standards: Memory loading in temporal reference memory. *Quarterly Journal of Experimental Psychology.*
Jones, L. A., & Wearden, J. H. (in press-b). More is not necessarily better: Examining the nature of the temporal reference memory component in timing. *Quarterly Journal of Experimental Psychology.*
Lejeune, H., Ferrara, A., Simons, F., & Wearden, J. H. (1997). Adjusting to changes in the time of reinforcement: Peak interval transitions in rats. *Journal of Experimental Psychology: Animal Behavior Processes*, *23*, 311–331.
Macar, F., Grondin, S., & Casini, L. (1994). Controlled attention sharing influences time estimation. *Memory and Cognition*, *22*, 673–686.

Matell, M. S., & Meck, W. H. (2000). Neuropsychological mechanisms of interval timing behavior. *Bioessays*, *22*, 94–103.
Meck, W. H. (1983). Selective adjustment of the speed of internal clock and memory processes. *Journal of Experimental Psychology: Animal Behavior Processes*, *9*, 171–201.
Penton-Voak, I. S., Edwards, H., Percival, A., & Wearden, J. H. (1996). Speeding up an internal clock in humans? Effects of click trains on subjective duration. *Journal of Experimental Psychology: Animal Behavior Processes*, *22*, 307–320.
Spetch, M. L., & Wilkie, D. M. (1983). Subjective shortening: A model of pigeons' memory for event duration. *Journal of Experimental Psychology: Animal Behavior Processes*, 9, 14–30.
Staddon, J. E. R., & Higa, J. J. (1999). Time and memory: Towards a pacemaker-free theory of interval timing. *Journal of the Experimental Analysis of Behavior*, *71*, 215–251.
Treisman, M., Faulkner, A., Naish, P. L. N., & Brogan, D. (1990). The internal clock: Evidence for a temporal oscillator underlying time perception with some estimates of its characteristic frequency. *Perception*, *19*, 705–748.
Wearden, J. H. (1991a). Do humans possess an internal clock with scalar timing properties? *Learning and Motivation*, *22*, 59–83.
Wearden, J. H. (1991b). Human performance on an analogue of an interval bisection task. *Quarterly Journal of Experimental Psychology*, *43B*, 59–81.
Wearden, J. H. (1992). Temporal generalization in humans. *Journal of Experimental Psychology: Animal Behavior Processes*, *18*, 134–144.
Wearden, J. H. (1994). Prescriptions for models of biopsychological time. In M. Oaksford & G. Brown (Eds.), *Neurodynamics and Psychology* (pp. 215–236). London: Academic Press.
Wearden, J. H. (1995). Categorical scaling of stimulus duration by humans. *Journal of Experimental Psychology: Animal Behavior Processes, 21*, 318–330.
Wearden, J. H. (1999). "Beyond the fields we know... :" Exploring and developing scalar timing theory. *Behavioural Processes*, *45*, 3–21.
Wearden, J. H. (2001). Internal clocks and the representation of time. In C. Hoerl & T. McCormack (Eds.), *Time and memory: Issues in philosophy and psychology* (pp. 37–58). Oxford: Clarendon Press.
Wearden, J. H., & Bray, S. (2001). Scalar timing without reference memory: Episodic temporal generalization and bisection in humans. *Quarterly Journal of Experimental Psychology, 54B,* 289–310.
Wearden, J. H., & Doherty, M. F. (1995). Exploring and developing a connectionist model of animal timing: Peak procedure and fixed-interval simulations. *Journal of Experimental Psychology: Animal Behavior Processes, 21*, 99–115.
Wearden, J. H., & Ferrara, A. (1993). Subjective shortening in humans' memory for stimulus duration. *Quarterly Journal of Experimental Psychology, 46B*, 163–186.
Wearden, J. H., & Ferrara, A. (1995). Stimulus spacing effects in temporal bisection by humans. *Quarterly Journal of Experimental Psychology*, *48B*, 289–310.
Wearden, J. H., & Grindrod, R. (2003). Manipulating decision processes in the human scalar timing system. *Behavioural Processes*, *61,* 47–56
Wearden, J. H., & McShane, B. (1988). Interval production as an analogue of the peak procedure: Evidence for similarity of human and animal timing processes. *Quarterly Journal of Experimental Psychology*, *40B*, 363–375.
Wearden, J. H., & Penton-Voak, I.S. (1995). Feeling the heat: Body temperature and the rate of subjective time, revisited. *Quarterly Journal of Experimental Psychology, 48B*, 129–141.

Wearden, J. H., Bajic, K., & Brocki, J. (submitted). Temporal bisection in humans: Response strategies, biases, and the location of the bisection point. *Journal of Experimental Psychology: Human Perception and Performance.*

Wearden, J. H., Denovan. L., Fakhri, M., & Haworth, R. (1997b). Scalar timing in temporal generalization in humans with longer stimulus durations. *Journal of Experimental Psychology: Animal Behavior Processes*, *23*, 502–511.

Wearden, J. H., Edwards, H., Fakhri, M., & Percival, A. (1998). Why "sounds are judged longer than lights:" Application of a model of the internal clock in humans. *Quarterly Journal of Experimental Psychology*, *51B*, 97–120.

Wearden, J. H., Parry, A., & Stamp, L. (2002). Is subjective shortening in human memory unique to time representations? *Quarterly Journal of Experimental Psychology, 55B,* 1–25.

Wearden, J. H., Wearden, A. J., & Rabbitt, P. (1997a). Age and IQ effects on stimulus and response timing. *Journal of Experimental Psychology: Human Perception and Performance*, *23*, 962–979.

Wing, A. M., & Kristofferson, A. B. (1973). The timing of interresponse intervals. *Perception and Psychophysics*, *20*, 191–197.

Chapter 3:
Psychological timing without a timer: The roles of attention and memory[*]

Richard A. Block

Abstract

Scalar-timing models have dominated explanations of animal timing. Scalar-timing theorists explain interval timing by assuming an internal-clock mechanism. For several reasons, this kind of model, along with other timing-with-a-timer models, may not provide a necessary and sufficient account of timing behavior and time experiences. Investigations of various human temporal judgments reveal influences of factors that may not be parsimoniously subsumed in an internal-clock framework. Evidence suggests that people make recency and serial position judgments by relying on both the apparent age of a past event (distance-based-processes) and on contextual associations to past events (location-based processes). When ongoing duration timing becomes important and salient, as in the human prospective duration judgment paradigm and the analogous animal paradigms (e.g., the peak procedure), attentional allocation becomes an important additional variable. I describe a memory-age model of processes involved in attending to time, which applies to relatively short-duration prospective timing. In retrospective timing, people apparently judge relatively long durations by relying mainly on availability of events and contextual changes associated with them. Timing-without-a timer models of psychological time (i.e., pacemaker-free models, such as the present memory-age model) need to be tested so that they can be evaluated against timing-with-a timer-models. I briefly review some neuropsychological evidence on temporal perspective, which involves remembering the past, experiencing the present, and anticipating the future. Researchers should consider whether timing-with-a-timer models adequately explain the many influences of attention and memory on duration experiences, as well as on the human ability to maintain a normal temporal perspective.

Introduction

Some of the first psychologists discussed and investigated the ways in which animals, including humans, experience and estimate time. Animals encode temporal aspects of stimuli and durations, remember those temporal aspects, and use stored temporal information to perform adaptive actions. In addition, humans (and perhaps

[*] I thank Hannes Eisler, Simon Grondin, Françoise Macar, John Moore, and Dan Zakay for helpful comments on a previous draft.

other, more advanced animals) orient themselves in time, remembering past events, experiencing present events, and planning or imagining future events. For more than a century, researchers have investigated the mechanisms or processes by which the mind solves problems concerning time. This research has produced abundant and diverse findings, as well as equally abundant and diverse theories.

Models of psychological time include those that assume one or more timer (internal clock, or pacemaker), and those that do not. Research focusing on the former kind of model, which is called *timing with a timer*, often differs considerably from research focusing on the latter kind of theory, which is called *timing without a timer* (Block, 1990; Ivry & Hazeltine, 1992). The two kinds of theorists often focus on different species, speak different languages, and use different paradigms. Many, although certainly not all, researchers studying timing with a timer focus on nonhuman animals (e.g., rats and pigeons), describe findings in operant conditioning terminology, and use a few relatively simple paradigms. Many, although certainly not all, researchers studying timing without a timer focus on humans, describe findings in cognitive terminology, and use diverse and relatively complex paradigms.

My early theorizing was in the cognitive tradition, emphasizing timing without a timer. I criticized internal clock models and discussed ways in which they are limited (Block, 1990). However, in the first *Time and Mind* volume (Helfrich, 1996), I proposed (along with Dan Zakay) a timing-with-a-timer model, the so-called *attentional-gate model* (Block & Zakay, 1996; see also Zakay & Block, 1996, 1997). This model represents a recent exception to the general characterization of timing-with-a-timer models as being focused on simple operant conditioning paradigms involving nonhuman animals.

I will argue here that timing-with-a-timer models do not provide a necessary and sufficient account of all aspects of psychological time. I will first briefly describe the most successful and widespread kind of model of timing with a timer, scalar-timing models. I will discuss some of their limitations. Finally, I will make the case for a class of models of timing without a timer by connecting psychological time with cognitive findings on attention and memory.

Timing with a timer: Scalar-timing models

Scalar-expectancy theory is an associative model of learning that is closely related to scalar-timing models, a class of psychophysical models of psychological time. Here, I will discuss them as if they were the same. Scalar-timing researchers have proposed several modules, including a pacemaker, a switch, an accumulator, a working memory, a reference memory, and a comparator (see Figure 1). Two previous authors (Church, this volume, chapter 1; Wearden, this volume, chapter 2) outlined some details of these models and the many findings they can explain, so I will summarize them here. The typical finding is that the psychophysical function relating physical duration and psychological duration is approximately linear. This is explained in terms of a pacemaker producing pulses at a fairly constant rate (with Poisson variability) and the accumulation of the pulses in working memory as a linear function of physical time. The typical finding that Weber's Law holds over a fairly wide range of durations is explained in terms of the process used in making the time estimate, which involves comparing the pulse total in working memory and the stored total in reference memory.

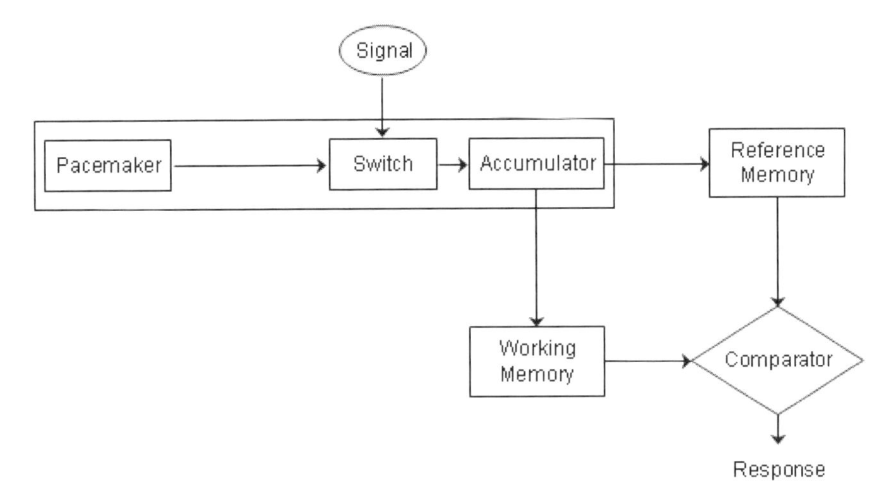

Figure 1: A typical scalar-timing model. Adapted from Gibbon, Church and Meck (1984, p. 54)[1]

Scalar-timing models have several drawbacks, or limitations:

1. No constant-rate pacemaker has been identified in the brain. This is perhaps not a serious problem, because researchers may ultimately find a neural basis for the pacemaker. Although some researchers have claimed to do so, this evidence is relatively weak (for a recent review, however, see Gibbon, Malapani, Dale, & Gallistel, 1997). Researchers have already revealed a neural basis for the pacemaker that underlies circadian rhythms, and they have shown that neurons in the suprachiasmatic nucleus can function endogenously, even when disconnected from normal inputs from nonvisual retinal photoreceptors (Freedman et al., 1999). Evidence that a pacemaker subserves interval timing on the order of seconds and minutes has, however, remained elusive.

2. Researchers advocating scalar-timing models have mainly used only a few paradigms, such as the peak procedure and the bisection task (see Wearden, this volume, chapter 2). In addition, until relatively recently, scalar-timing researchers have mainly studied rats and pigeons. Although some scalar-timing researchers have recently investigated counting, foraging, and other animal behavior, as well human timing behavior (e.g., Wearden, this volume, chapter 2; Wearden & Culpin, 1998), most of the evidence comes from a few relatively simple paradigms.

---

[1] Reprinted with permission.

3. Perhaps because it is relatively difficult to devise situations in which nonhuman animals must make various kinds of temporal judgments, the evidence supporting scalar-timing models has come mainly from studies in which animals estimate the duration of a single stimulus or an interval between two stimuli (see, however, Church, this volume, chapter 1). Most scalar timing experiments have used so-called *empty* durations, or gaps between stimuli, during which no external stimuli are presented, and the animal has no external information to process. Scalar-timing models cannot easily explain a wide variety of human research using so-called *filled* durations. The human literature on psychological time, which I will discuss next, also includes various other kinds of judgments, such as temporal location and recency judgments. Scalar-timing models were not designed to explain these kinds of judgments, and they cannot easily be adapted to do so.

4. Scalar-timing models are not easily able to explain effects of attention on psychological time, which are widespread and well documented in the human cognitive literature (see later). Most discussions of the role of attention in scalar timing are relatively brief, usually attributing attentional effects to processes operating on the proposed switch, such as simple delays in closing it (see, however, Lejeune, 1998).

5. Many of the findings that scalar-timing models explain are generic. For example, the basic notion—the so-called *scalar property* of timing—is found in psychophysical judgments involving a wide variety of physical dimensions for which a ratio scale is appropriate (Eisler, 1965). The finding that the standard deviation of estimates increases proportionally with the mean estimate is typical for many dimensions, and the approximate constancy of the coefficient of variation (standard deviation divided by mean estimate), along with the closely related Weber's Law, is not unique to the time dimension. Only the pacemaker and accumulator components are unique to the time dimension. With only slight modification (e.g., substituting external stimulus information for the pacemaker), scalar-timing models could easily become scalar-perceiving models.

6. The typical scalar-timing assumption that time estimates are a linear function of physical duration is not widely supported. Eisler (1976), for example, reported that psychological time is a power function of physical time, with an exponent less than 1.0 (about 0.9). More recently, Staddon and Higa (1999) suggested that psychological time is a logarithmic function of physical time, which they explained in terms of a process of habituation.

Timing without a timer: Attention and memory models

In the rest of this chapter, I will lay some of the groundwork for theories that do a better job of integrating findings concerning psychological time with well-established findings concerning attention and memory. I will first distinguish studies that focused on when a past stimulus (event) occurred, how long a single stimulus lasted, and how long an interval lasted.

Temporal location judgments

Two separate processes apparently subserve memory for the times (temporal locations) of past events, which are usually called *distance-based processes* and *location-based processes* (for a recent review, see Friedman, 2001). Distance-based processes depend on the continually changing amount of time that has elapsed between some past event and the present, whereas location-based processes depend on the relationships between some past event and relatively unchanging memories for past time patterns. Neither distance-based processes nor location-based processes requires any sort of pacemaker-accumulator system or any *special* encoding of time. Distance-based processes involve a judgment of the vividness of the memory for the event. Location-based processes involve inferences about other events in which it was embedded. Consider some of the evidence.

Distance-based processes. Surprisingly little evidence requires an explanation in terms of distance-based processes. Nevertheless, there is some. Friedman (1991), for example, found that 4- to 8-year-old children could accurately remember the relative recency of events, one that they had experienced one week earlier and another that they had experienced seven weeks earlier. However, the children could not remember the day, month, or season during which each event had occurred. Friedman (2001) argued that the children based their memory for the time of a past event on an impression of its age, not on a process of remembering the location of the event in a pattern of events. Friedman and Kemp (1998) found that this impressionistic information is a decelerating function of the actual age of the event, with most of the change occurring during the preceding few months (see Figure 2). The best-fitting power function had an exponent of only 0.20.

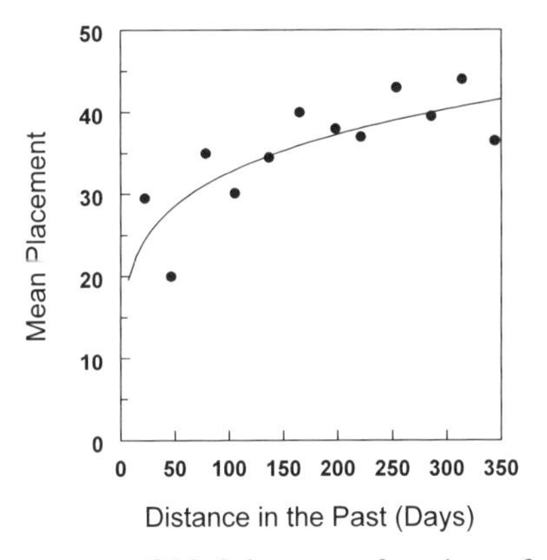

Figure 2: Mean placement of birthday as a function of actual distance in the past. Adapted from Friedman and Kemp (1998, p. 357)[2]

[2] Reprinted with permission from Elsevier.

Location-based processes. Location-based processes are probably more important and more accurate than distance-based processes for human adults. These depend mainly on the relatively automatic encoding of events in a rich cognitive context, along with inferences at the time of retrieval and judgment. In one experiment (Hintzman & Block, 1971) participants viewed a series of 50 words without being forewarned that temporal judgments would be required. They were subsequently asked to judge the temporal position of each word in the series. If they remembered the word, they were able to judge its temporal position with some accuracy (see Figure 3), especially those near a landmark event, the start of the word series. In subsequent experiments, participants viewed words during two separate durations, again under incidental conditions (Hintzman, Block, & Summers, 1973). Afterwards, they were asked to judge whether each word had occurred during the first or the second duration, and then whether it had occurred near the beginning, middle, or end of it. They were able to make these judgments with some accuracy. However, if a participant incorrectly judged that a word had occurred during a particular duration, he or she nevertheless tended to judge that it had occurred in the correct part of it. Thus, remembered events did not randomly migrate to temporally adjacent locations. These location-based temporal position judgments apparently rely on incidentally (i.e., relatively automatically) encoded contextual information. As such, contextual information enables people to locate events on a relative scale of psychological time, not on a continuous scale of absolute time (cf. Hintzman, 2001, 2002).

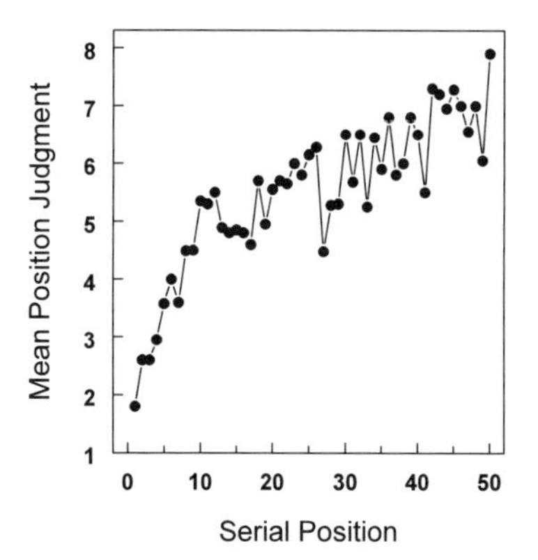

Figure 3: Mean judgment of temporal position judgment (on a scale from *1* to *10*) as a function of the serial position of a word in a series of words. Adapted from Hintzman and Block (1971, p. 299)[3]

Similar evidence comes from studies of autobiographical memory in which participants are asked to date personal memories of events that occurred relatively long ago, such as months or years in the past. Participants can temporally locate their personal memories with some accuracy. However, their judgments also reveal systematic errors called *scale effects*. For example, an event may be accurately remembered as having occurred during a particular time of day, but inaccurately remembered as to the day, month, or year (Friedman & Wilkins, 1985). Thus, recency judgments involve location-based processes, not just distance-based processes, and important contextual landmarks influence these processes (Friedman, 1996; Shum, 1998).

In short, evidence shows clearly that normal memory and cognitive processes are necessary and sufficient to explain memory for the time of a past event. The information is encoded relatively automatically, and there is no need to assume an internal-clock mechanism, such as the pacemaker-accumulator system of scalar timing models. Indeed, if one were to assume such a mechanism, one would need to postulate that a separate internal clock is switched on for each experienced event. The brain undoubtedly does not contain a separate pacemaker and accumulator for each event that is experienced. This has important implications for studies of interval timing in animals and humans, which I now discuss.

Stimulus timing

Theories of timing must distinguish between two kinds of duration judgments: judging the duration of a single stimulus (or the relative duration of two stimuli), and judging the interval between two events. These two kinds of time judgments almost certainly involve different processes.

Animals can learn to make one response to a relatively short-duration stimulus and another response to a relatively long-duration stimulus (Fetterman, 1995). If they are tested on novel stimuli presented at intermediate durations, they usually bisect (that is, show mathematical indifference) at the geometric mean of the two learned stimuli. If a novel test stimulus is presented at an intermediate duration and a delay is interposed between it and the time the animal is permitted to respond, the animal tends to remember that the test stimulus is shorter than if no delay had been interposed. This reliable finding, called the *choose-short effect*, seems to indicate that the animal has forgotten some of the temporal information. Although a scalar-timing model could assume that some pulses are lost from the accumulator over time, evidence suggests that the choose-short effect is better explained in terms of proactive interference (stimulus generalization) or other well-known memory phenomena (Kraemer, Mazmanian, & Roberts, 1985). People can also remember the approximate duration of each event in a long series of events, and they can do so even if they were not expecting to perform the task (Hintzman, 1970). Duration information is apparently encoded relatively automatically as an integral part of the experience of an event.

Interval timing

Scalar-timing models cannot easily explain all the findings concerning judgments of the duration of a stimulus. Consider now whether scalar-timing models can explain all the findings concerning judgments of intervals between events. Many researchers have investigated the processes involved in judging the duration of a relatively long empty interval (i.e., one containing no changes in external stimuli) or judging the duration of a relatively long filled interval (i.e., one containing changes in external stimuli, such as a series of stimulus events). In these cases, duration is not a property of a single stimulus event.

Scalar-timing models were originally proposed to explain interval timing—and *only* interval timing. They have successfully explained and guided much research on interval timing. In one common paradigm, called the *peak procedure*, rats or pigeons learn to expect reinforcement after a fixed interval (usually on the order of tens of seconds), and during this interval no external stimuli are presented. Because it is not uncommon for an animal to begin responding after only about one-third of the required interval has elapsed, one could argue that one or more proposed modules (pacemaker, switch, accumulator, memory, and comparator) must operate in a rather imprecise way. The animals' difficulties are perhaps reflected in the common finding that stereotyped behavior chains (so-called *adjunctive behaviors*) occur during the interval (Killeen & Fetterman, 1988). These adjunctive behaviors may reflect an adaptive strategy to provide timing in the absence of an accurate internal clock.

Scalar-timing proponents have investigated interval timing under limited conditions, at least until relatively recently. Typically, an animal is observed behaving during an empty interval, one during which no changing external stimuli are presented or processed. Although some scalar-timing researchers have recently been investigating more ecologically valid conditions, with changing external stimulation during the interval (e.g., Lejeune, Macar, & Zakay, 1999), most of what is known about interval timing comes from human research.

Human researchers investigating interval (duration) timing work mainly in a cognitive tradition, and they study effects of varying information-processing tasks that a person must perform during a time period. Duration timing reveals interactions among conditions prevailing when a time period is experienced and those prevailing when it is judged (Block, 1989). For example, different findings are sometimes obtained depending on the choice of duration-estimation task. The usual tasks involve production, verbal estimation, reproduction, and similar methods (for a review, see Zakay, 1990).

Prospective timing. The method most analogous to the peak procedure in animal timing is production, in which a person is asked to delimit an objective interval to estimate a verbally stated duration, such as "30 seconds." People make such estimates under what is called the *prospective paradigm*, in which they are aware that timing is relevant and important. I refer to prospective duration judgments as reflecting *experienced duration*. If an experimenter requires a participant to perform other information-processing tasks during the production, experienced duration varies along with these attention-demanding processes. If there are fewer stimuli or if a processing task is easy, experienced duration increases, as revealed by shorter productions or larger verbal estimates of duration (Hicks, Miller, & Kinsbourne, 1976; Zakay, 1993; Zakay & Block, 1997). Prospective timing is therefore a dual-task condition in which attention is shared between nontemporal and temporal information processing. Nontemporal information processing is directed toward external stimuli (along with accompanying internal cognitions), excluding attributes involving time. Temporal information processing is directed toward time-related aspects of external stimuli, as well as time-related internal cognitions (such as what is called *attending to time*). Many findings reveal that temporal information processing requires access to some of the same attentional resources that attending to nontemporal information does. Experienced duration increases if a person allocates relatively more attentional resources to processing temporal information. If a person is told how much attention to allocate for stimulus information processing and how much to allocate for temporal information processing, prospective duration judgments depend on the relative allocation (Brown, 1997; Macar, Grondin, & Casini, 1994; Zakay, 1992, 1998). If a person must track the duration of several concurrent events, timing accuracy decreases as a function of the number of monitored events (Brown, 1997).

For this reason, most theorists emphasize the role of attentional resource allocation (Block & Zakay, 1996; Brown, 1998; Macar et al., 1994; Thomas & Weaver, 1975; Zakay & Block, 1996). Some scalar-timing theorists have said that attention is needed to operate a switch between the pacemaker and the accumulator (Meck, 1984). Until recently, the attentional effect on the switch was limited to the requirement that the animal perceives the signal indicating that the interval had begun. Other than that attentional requirement, the typical scalar-timing model did not incorporate attentional effects in any serious way (see, however, Lejeune, 1998). This lead Zakay and me (Block & Zakay, 1996; Zakay & Block, 1996, 1997) to propose what we called an *attentional-gate model* of prospective duration judgments (see Figure 4). The main difference between this model and scalar-timing models is that an attentional gate is interposed between the pacemaker and the accumulator, and this attentional gate allows pulses produced by a pacemaker to be accumulated only when it is operated by attention. Lejeune (1998) questioned the need to propose both a switch and a gate, but there is a major theoretical difference between attending to the duration-onset signal and attending to time during the duration (see Zakay, 2000).

Scalar-timing models and the attentional-gate model are both pacemaker-accumulator systems. If one does not adopt the assumption that a pacemaker-accumulator system underlies prospective timing, what is the alternative? One possibility is that interval timing involves a comparing apparent ages of events. Assume that the apparent age of an event (which is the inverse of apparent recency) increases as a negatively accelerated (e.g., power) function of physical time, as in the findings of Friedman and Kemp (1998; see the present Figure 2).

When a person is asked actively to produce a verbally stated duration, the person terminates the production when the apparent age of the start (duration-onset) signal matches the average apparent age for that approximate duration that has been learned in the past. Analogously, in the peak procedure the animal responds to the extent that the apparent age of the start signal matches the apparent age of the start signal at the time of reinforcement during previous trials. When a person is asked to verbally estimate a past duration, the relevant comparison involves the apparent ages of the start-of-duration and end-of-duration events in memory, and the person translates this information into numerical time units based on similar comparisons stored in the past. When a person is asked to reproduce a past duration, the person encodes the apparent age of the start signal at the time the end signal occurs and then terminates the reproduction (as in the method of production) when the apparent age of the start of the reproduction is comparable to it.

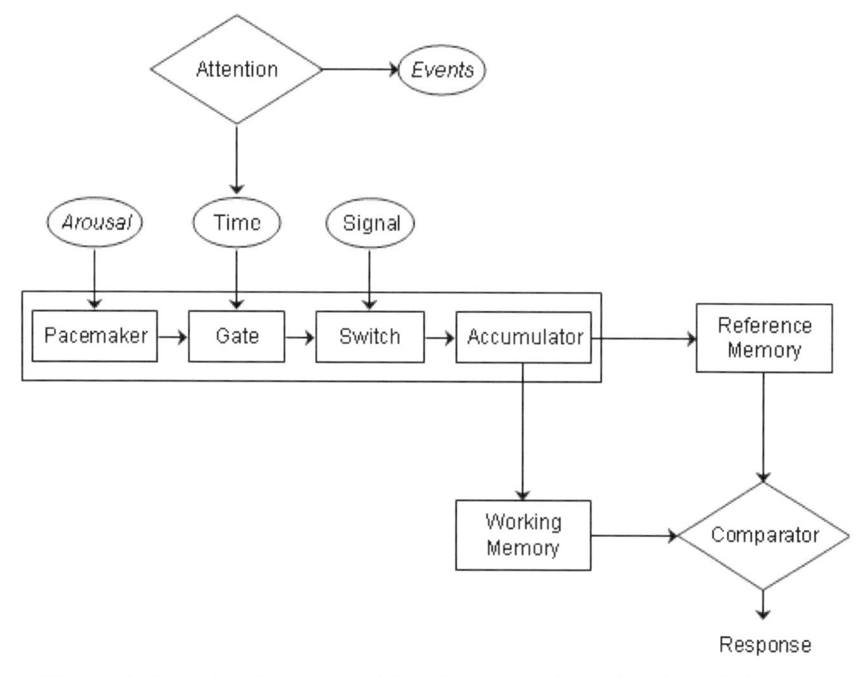

Figure 4: Attentional-gate model of prospective duration judgment. Adapted from Block and Zakay (1996, p. 182).[4]

If prospective duration timing involves comparing the relative ages of the start and end signals, the process by which information-processing demands during the duration influences the comparison needs to be clarified. If a person has few attentional demands during the time period, the typical explanation is that the person is able to attend to time more often and, as a result, stores more temporal information. What a person does when he or she attends to time has never been fully described. One possibility is that every act of attending to time involves retrieval of information concerning the apparent age of the previous act of attending to time. Because apparent age increases as a negatively accelerated (e.g., power) function of physical time, on every occasion that age information is retrieved, the accumulated age information increases in an unusually large way. In other words, the process involves accumulating samples of relatively large changes in relative age. If a person attends to time less often, or not at all, apparent-age information is only retrieved a few times, or not at all, and the power-function aging process is nearer to an asymptotic level. This

[4] Reprinted with permission from Hogrefe and Huber Publishers.

model, which I call a *memory-age model* of prospective duration timing, is a plausible alternative to timing-with-a-timer models of interval timing.

A process that underlies the retrieval of age information was originally called *study-phase retrieval*. It refers to the relatively automatic way in which information associated with an earlier event is retrieved by the same event or a similar event (Hintzman, Summers, & Block, 1975; Tzeng & Cotton, 1980). The retrieved information is contextual in nature, including information on apparent recency or age of the previous occurrence. Attending to time may automatically retrieve information about the previous act of attending to time, including the apparent age of that act. Repeated acts of attending to time may increase experienced duration by means of this retrieval process, and perhaps also by increasing the segmentation of the duration (see later).

Retrospective timing. In contrast to prospective duration judgments, in the retrospective paradigm people do not know or suspect that they will be asked to judge duration until after the time period has ended. The animal literature has not provided any experimental evidence on retrospective duration judgments; such experiments are difficult or perhaps impossible to undertake. Without any previous operant or respondent conditioning, which would lead the animal to learn that timing is relevant, the experimenter must communicate to the animal that it should estimate a *past* duration. The animal must understand the communication and have some minimal concept of time as involving both past and present. According to Tulving (2002), only humans display what he called *autonoetic consciousness*. This kind of consciousness enables a human to grasp concepts of past, present, and future, and this kind of temporal perspective may be required for retrospective duration estimation. A clever future experimenter may devise a way to ask an animal, perhaps a nonhuman primate, to judge a past duration, and comparative psychological evidence may become available in the future.

Many researchers have studied prospective duration judgments, but few have studied retrospective duration judgments (for a notable exception, see Eisler & Eisler, 2001). The main reason is that after a participant provides a retrospective judgment, he or she will suspect that additional duration judgments will be requested. Thus, the paradigm becomes prospective. Collecting data on retrospective duration judgments requires many participants, because ordinarily each participant can provide only one estimate. The exceptions involve retrospective duration estimates of autobiographical events, or perhaps of several durations presented in the laboratory, but there are serious difficulties in interpreting such evidence. Later, I will discuss time-order effects.

Several kinds of variables influence the magnitude of retrospective duration judgments, or what I call *remembered duration*. Remembered duration increases if a person processed and can now remember a greater number of stimuli or more complex stimuli (Ornstein, 1969). However, remembered duration is not based simply on memory of individual events or their encoded complexity (Block, 1974; Block & Reed, 1978). In addition, information-processing demands do not influence remembered duration much, if at all (Block & Zakay, 1997; Hicks et al., 1976). Instead, changes in context increase remembered duration. These contextual changes may involve environment stimuli, internal stimuli, information processing strategies, and other such factors (Block, 1989). For example, remembered duration increases to the extent that a person performed different kinds of tasks during the time period. It increases if a person has had no previous experience in a particular environment (Block, 1982). It also increases if the time period was segmented by high-priority events, such as politicians' names inserted among names of animals (Poynter, 1989). Such segmentation may cause contextual changes when the high-priority events appear.

Retrospective duration judgments usually show what is called a *positive time-order effect*: The first of two equal time periods is remembered as being longer than the second (Block, 1982, 1985). This effect is somewhat counterintuitive: Some models predicts the opposite, a negative time-order effect attributable to older memories dropping out of storage, or fading with time (Ornstein, 1969). However, a person may encode a greater number of contextual changes during a novel experience, such as during the first of two durations experienced in the laboratory. Several findings suggest that contextual changes underlie the positive time-order effect. If the environmental context prevailing during the second duration is different from that prevailing during the first, the effect is eliminated (Block, 1982). If changes in emotions or mood that would ordinarily occur during the first duration occur instead during a preceding time period, the effect is also eliminated (Block, 1986). If the interval between two durations increases, so that the first one becomes relatively less recent and the second one becomes relatively more recent, the effect may reverse, becoming a negative time-order effect (Wearden & Ferrara, 1993). In this case, people may have difficulty remembering the contextual changes that occurred during the first duration.

Scalar timing models were not developed to explain retrospective duration estimates, and they have difficulty doing so. One problem is that durations of event sequences often overlap. Consider a situation in which a person drives a car for 90 minutes. During a 50-minute duration near the start of the car trip, the person listens to some music on the car stereo. Partially overlapping with this duration, the person opens and drinks a bottle of water during a 35-minute duration. After the car journey, the person may be able to provide reasonable estimates of the total duration of the journey, the duration of the music, and the duration of the water drinking (although these estimates may show considerable variability, especially relative to prospective estimates). Several concurrently operating internal clocks would be needed to explain this ability. In real-world situations such as this, the number of concurrently operating internal clocks could easily proliferate to a large number. In addition, many timing-with-a-timer models cannot explain why many information-processing variables (such as number and complexity of events, segmentation of events, and contextual changes) influence retrospective timing.

Perhaps the same kind of cognitive model that can explain prospective timing can also explain retrospective timing. For example, a person may retrieve and compare the apparent ages of the event that signaled the start of the duration and the event that signaled the end of the duration. This seems unlikely. In order for this simple model to be viable, it would have to explain the process by which the nature of events and contextual changes that occurred during the duration influence the apparent age of the start-of-duration signal, the apparent age of the end-of-duration signal, or the comparison of the two.

Prospective versus retrospective timing. Evidence from experiments that compared prospective and retrospective duration estimates suggests that different processes are involved in the two paradigms. Prospective judgments are usually larger in magnitude and smaller in variability than are retrospective judgments, and several variables influence duration judgments differently in the two paradigms (Block & Zakay, 1997). For example, prospective verbal estimates decrease if a person had performed a relatively difficult information-processing task, but retrospective estimates are not affected (Block, 1992; Hicks et al., 1976). In addition, retrospective estimates increase if a person had performed different information-processing tasks, but prospective estimates are not usually affected (Block, 1992; but see Brown, 1997). These findings suggest that different processes subserve duration judgments in the two paradigms.

In the retrospective paradigm, people probably do not attend to time much unless there is little information to process or a boring situation. Consequently, most models emphasize that retrospective estimates must rely on some aspect of episodic memory for events that occurred during the duration. Theorists have proposed that remembered duration is based on the "multitudinousness of the memories which the time affords" (James, 1890); stored and retrieved information, or "storage size" (Ornstein, 1969); remembered changes (Fraisse, 1963); encoded and retrieved contextual changes (Block, 1974); interval segmentation (Poynter, 1983); and other such constructs. In the retrospective paradigm, people may selectively attempt to retrieve memories of some events that occurred during the time period. Remembered duration increases to the extent that the events are more easily retrievable. Thus, people may use an availability heuristic. However, remembered duration is not based entirely on availability to memory of external events. Instead, people apparently judge a duration based mainly on the contextual changes that were automatically encoded in memory along with the external events (Block, 1982; Block & Reed, 1978). If a person is able to retrieve memories for more external events at the time of the retrospective duration judgment, more contextual information is also retrieved, because contextual information is activated when a person remembers an event. In the prospective paradigm, these contextual changes are also automatically encoded, of course, but they apparently play a minor role. Instead, the person may rely mainly on the changes in the apparent age of the events associated with each act of attending to time.

To state it differently, both distance-based information and location-based information may be used to make prospective and retrospective duration judgments. However, the relative importance of the two kinds of information may differ. In prospective timing, distance-based information may be relatively more important. Distance-based information (i.e., apparent age of a past event) is sensitive over the short time periods that are usually involved in prospective timing (seconds to minutes). Distance-based information may be the main information available to animals and children. In retrospective timing, on the other hand, location-based information may be relatively more important. Location-based information (i.e., contextual associations to events) is sensitive over the long time periods that are usually involved in retrospective timing (minutes, hours, days, weeks, or years). Location-based information may be the main kind of information that adult humans use to estimate relatively long time periods, especially those that occurred relatively long ago.

Temporal perspective

Most researchers investigating psychological time have focused on a person's ability to estimate when a past event occurred (recency or temporal position judgments), how long a past stimulus lasted (stimulus-duration judgments), or how long a passing or past series of events lasted (prospective or retrospective duration judgments). Another aspect of psychological time involves what is usually called *temporal perspective* (Block, 1979). Involuntary and voluntary shifts of attention to past, present, and future events change the contents of consciousness. At any moment, a person may be remembering a dinner conversation last night, focusing on what someone is saying now, or thinking about what to order for dinner tonight. Even though focal attention is different in these three cases, a person usually maintains awareness of the present context. As I have already noted, contextual elements seem to be automatically associated with events, whether the event is mainly externally triggered (as in perceiving) or mainly internally generated (as in remembering or planning). A person may be able to remember that he or she (a) last thought about the previous night's dinner conversation about 40 seconds ago, (b) read the word *contextual* about 15 seconds ago, and (c) thought about tonight's dinner plans about 50 minutes ago. Although focal attention and corresponding awareness may be oriented toward the past (remembering), toward the present (perceiving), or toward the future (planning), the present context is always just outside of focal attention, and it becomes associated with the remembering, perceiving, or planning activity. This relatively automatic construction of context may give rise to the apparent continuity of our consciousness in spite of shifts from remembering the past to perceiving the present to planning the future.

Sometimes this does not occur. The literature contains descriptions of experiences of timelessness, which may accompany creative states, meditative states, or psychoactive drug-induced states of consciousness, among others (Block, 1979). Some people have reported wakening in a strange hotel room and not knowing, for a few seconds, where or when they are—that is, in what city they are, at what time of day it is, or even what day it is. This is reminiscent of the description of H.M., a patient who received a bilateral hippocampectomy. As a result of his inability to form new long-term episodic memories, H.M. lives in a present that extends only about 15 seconds back into the past. He said that his continual mental condition is "like waking from a dream" (Milner, 1970, p. 37). Although in people with an intact hippocampus this experience is usually brief, a typical description is that one loses the ordinary impression of time and place. Perhaps, for some unknown reason (probably involving the functioning of the frontal lobes, which may construct context and send contextual information to the hippocampus), the ordinarily automatic maintenance of the present cognitive context ceases to occur. H.M., whose frontal lobes were intact, could engage in planning, although he may not have been able to remember his plan if more than about 15 seconds elapsed between the planning and the opportunity to engage in the planned activity.

Tulving, along with his colleagues (Wheeler, Stuss, & Tulving, 1997), has recently discussed what he calls *chronesthesia*, "a form of consciousness that allows individuals to think about the subjective time in which they live and that makes it possible for them to 'mentally travel' in such time" (Tulving, 2002, p. 311). He argued that only humans older than about three or four years of age have sufficiently well developed frontal lobes to subserve chronesthesia. He also reviewed evidence on K.C., a neurological patient with multiple cortical and white matter lesions in both anterior and posterior parts of the brain, along with hippocampal damage. Like H.M., K.C. has no functional episodic memory: He cannot remember anything that happened to him personally more than about 15 seconds ago. Also like H.M., his working memory is intact, and he is an intelligent person. However, K.C. has little or no concept of his personal future. When he is asked to think about the next half hour or the next year, he says that his mind is "blank." Tulving (2002) concluded: "K.C. seems to be as incapable of imagining his future as he is of remembering his past" (p. 317). However, K.C. *knows* about time: He knows about clocks and calendars, and he can talk about what he and other people know about physical time, including what day it is, and so on.

In contrast to H.M. and K.C., who know about impersonal time, some patients with lesions of the dorsomedial nucleus of the thalamus show an impairment of temporal orientation that has been called *chronotaraxis* (Spiegel, Wycis, Orchinik, & Freed, 1955). In chronotaraxis, the patient is unable to know the season, the date, the day of week, or the time of day.

Models of psychological time, if they are to be complete and integrative models, will eventually have to include what researchers are beginning to discover about temporal perspective. Future, more integrative models will need to include awareness of present time and future time with awareness of the ages of past events, awareness of past event durations, and awareness of the durations of passing or past sequence of events.

Conclusion

Pacemaker-accumulator models of timing, such as scalar-timing models, are limited in scope. These models sufficiently account for some findings, especially on animal timing of stimuli and intervals, but they may not be necessary. Timing-with-a-timer models were originally devised to explain only interval timing, and they cannot easily explain other aspects of psychological time, such as remembering the approximate age of events, making retrospective duration judgments, and maintaining a temporal perspective.

Pacemaker-free models, on the other hand, provide a necessary and sufficient account of memory for stimulus duration and interval timing, as well as the other aspects of psychological time that I have discussed. They do so in a potentially integrative way, mainly by focusing on the role of attention and memory processes. Pacemaker-free models have not yet been developed to provide precise, mathematical predictions. However, in the near future they may be able to do so. Further exploration of pacemaker-free cognitive models of psychological time is needed.

References

Block, R. A. (1974). Memory and the experience of duration in retrospect. *Memory and Cognition*, *2*, 153–160.

Block, R. A. (1979). Time and consciousness. In G. Underwood & R. Stevens (Eds.), *Aspects of consciousness: Vol. 1. Psychological issues* (pp. 179–217). London: Academic Press.

Block, R. A. (1982). Temporal judgments and contextual change. *Journal of Experimental Psychology: Learning, Memory, and Cognition*, *8*, 530–544.

Block, R. A. (1985). Contextual coding in memory: Studies of remembered duration. In J. A. Michon & J. L. Jackson (Eds.), *Time, mind, and behavior* (pp. 169–178). Berlin: Springer.

Block, R. A. (1986). Remembered duration: Imagery processes and contextual encoding. *Acta Psychologica*, *62*, 103–122.

Block, R. A. (1989). Experiencing and remembering time: Affordances, context, and cognition. In I. Levin & D. Zakay (Eds.), *Time and human cognition: A life-span perspective* (pp. 333–363). Amsterdam: North-Holland.

Block, R. A. (1990). Models of psychological time. In R. A. Block (Ed.), *Cognitive models of psychological time* (pp. 1–35). Hillsdale, NJ: Erlbaum.

Block, R. A. (1992). Prospective and retrospective duration judgment: The role of information processing and memory. In F. Macar, V. Pouthas, & W. J. Friedman (Eds.), *Time, action and cognition: Towards bridging the gap* (pp. 141–152). Dordrecht, The Netherlands: Kluwer Academic.

Block, R. A., & Reed, M. A. (1978). Remembered duration: Evidence for a contextual-change hypothesis. *Journal of Experimental Psychology: Human Learning and Memory*, *4*, 656–665.

Block, R. A., & Zakay, D. (1996). Models of psychological time revisited. In H. Helfrich (Ed.), *Time and mind* (pp. 171–195). Seattle, WA: Hogrefe & Huber Publishers.

Block, R. A., & Zakay, D. (1997). Prospective and retrospective duration judgments: A meta-analytic review. *Psychonomic Bulletin and Review*, *4*, 184–197.

Brown, S. W. (1997). Attentional resources in timing: Interference effects in concurrent temporal and nontemporal working memory tasks. *Perception and Psychophysics*, *59*, 1118–1140.

Brown, S. W. (1998). Automaticity versus timesharing in timing and tracking dual-task performance. *Psychological Research/ Psychologische Forschung*, *61*, 71–81.

Eisler, H. (1965). The connection between magnitude and discrimination scales and direct and indirect scaling methods. *Psychometrika*, *30*, 271–289.

Eisler, H. (1976). Experiments on subjective duration 1868–1975: A collection of power function exponents. *Psychological Bulletin*, *83*, 1154–1171.

Eisler, A. D., & Eisler, H. (2001). A quantitative model for retrospective subjective duration. *Manuscript submitted for publication.*

Fetterman, J. G. (1995). The psychophysics of remembered duration. *Animal Learning and Behavior*, *23*, 49–62.

Fraisse, P. (1963). *The psychology of time* (J. Leith, Trans.). New York: Harper & Row. (Original work published 1957).

Freedman, M. S., Lucas, R. J., Soni, B., von Schantz, M., Munoz, M., David-Gray, Z., et al. (1999). Regulation of mammalian circadian behavior by non-rod, non-cone, ocular photoreceptors. *Science*, *284*, 502–504.

Friedman, W. J. (1991). The development of children's memory for the time of past events. *Child Development*, *62*, 139–155.

Friedman, W. J. (1996). Distance and location processes in memory for the times of past events. *Psychology of Learning and Motivation*, *35*, 1–41.

Friedman, W. J. (2001). Memory processes underlying humans' chronological sense of the past. In C. Hoerl & T. McCormack (Eds.), *Time and memory: Issues in philosophy and psychology* (pp. 139–167). Oxford, England: Oxford University Press.

Friedman, W. J., & Kemp, S. (1998). The effect of elapsed time and retrieval of young children's judgments of the temporal distances of past events. *Cognitive Development*, *13*, 335–367.

Friedman, W. J., & Wilkins, A. J. (1985). Scale effects in memory for the time of events. *Memory and Cognition*, *13*, 168–175.

Gibbon, J., Church, R. M., & Meck, W. (1984). Scalar timing in memory. In J. Gibbon and L. Allan (Eds.), *Annals of the New York Academy of Sciences*, *423: Timing and time perception* (pp. 52–77). New York: New York Academy of Sciences.

Gibbon, J., Malapani, C., Dale, C. L., & Gallistel, C. R. (1997). Toward a neurobiology of temporal cognition: Advances and challenges. *Current Opinion in Neurobiology*, *7*, 170–184.

Helfrich, H. (Ed.). (1996). *Time and mind.* Seattle, WA: Hogrefe & Huber Publishers.

Hicks, R. E., Miller, G. W., & Kinsbourne, M. (1976). Prospective and retrospective judgments of time as a function of amount of information processed. *American Journal of Psychology*, *89*, 719–730.

Hintzman, D. L. (1970). Effects of repetition and exposure duration on memory. *Journal of Experimental Psychology*, *83*, 435–444.

Hintzman, D. L. (2001). Judgments of frequency and recency: How they relate to reports of subjective awareness. *Journal of Experimental Psychology: Learning, Memory, and Cognition*, *27*, 1347–1358.

Hintzman, D. L. (2002). Context matching and judgments of recency. *Psychonomic Bulletin and Review*, *9*, 368–374.

Hintzman, D. L., & Block, R. A. (1971). Repetition and memory: Evidence for a multiple-trace hypothesis. *Journal of Experimental Psychology*, *88*, 297–306.

Hintzman, D. L., Block, R. A., & Summers, J. J. (1973). Contextual associations and memory for serial position. *Journal of Experimental Psychology*, *97*, 220–229.

Hintzman, D. L., Summers, J. J., & Block, R. A. (1975). Spacing judgments as an index of study-phase retrieval. *Journal of Experimental Psychology: Human Learning and Memory*, *1*, 31–40.

Ivry, R. B., & Hazeltine, R. E. (1992). Models of timing-with-a-timer. In F. Macar, V. Pouthas, & W. J. Friedman (Eds.), *Time, action and cognition: Towards bridging the gap* (pp. 183–189). Dordrecht, The Netherlands: Kluwer Academic.

James, W. (1890). *The principles of psychology.* (Vol. 1). New York: Holt.

Killeen, P. R., & Fetterman, J. G. (1988). A behavioral theory of timing. *Psychological Review, 95*, 274–295.

Kraemer, P. J., Mazmanian, D. S., & Roberts, W. A. (1985). The choose-short effect in pigeon memory for stimulus duration: Subjective shortening versus coding models. *Animal Learning and Behavior, 13*, 349–354.

Lejeune, H. (1998). Switching or gating? The attentional challenge in cognitive models of psychological time. *Behavioural Processes, 44*, 127–145.

Lejeune, H., Macar, F., & Zakay, D. (1999). Attention and timing: Dual-task performance in pigeons. *Behavioural Processes, 45*, 141–157.

Macar, F., Grondin, S., & Casini, L. (1994). Controlled attention sharing influences time estimation. *Memory and Cognition, 22*, 673–686.

Meck, W. H. (1984). Attentional bias between modalities: Effect on the internal clock, memory, and decision stages used in animal time discrimination. In J. Gibbon & L. G. Allan (Eds.), *Annals of the New York Academy of Sciences: Vol. 423. Timing and time perception* (pp. 528–541). New York: New York Academy of Sciences.

Milner, B. (1970). Memory and the medial temporal regions of the brain. In K. H. Pribram & D. E. Broadbent (Eds.), *Biology of memory* (pp. 29–50). New York: Academic Press.

Ornstein, R. E. (1969). *On the experience of time*. Harmondsworth, England: Penguin.

Poynter, W. D. (1983). Duration judgment and the segmentation of experience. *Memory and Cognition, 11*, 77–82.

Poynter, W. D. (1989). Judging the duration of time intervals: A process of remembering segments of experience. In I. Levin & D. Zakay (Eds.), *Time and human cognition: A life-span perspective* (pp. 305–331). Amsterdam: North-Holland.

Shum, M. S. (1998). The role of temporal landmarks in autobiographical memory processes. *Psychological Bulletin, 124*, 423–442.

Spiegel, E. A., Wycis, H. T., Orchinik, C. W., & Freed, H. (1955). The thalamus and temporal orientation. *Science, 121*, 771–772.

Staddon, J. E. R., & Higa, J. J. (1999). Time and memory: Towards a pacemaker-free theory of interval timing. *Journal of the Experimental Analysis of Behavior, 71*, 215–251.

Thomas, E. A., & Weaver, W. B. (1975). Cognitive processing and time perception. *Perception and Psychophysics, 17*, 363–367.

Tulving, E. (2002). Chronesthesia: Conscious awareness of subjective time. In D. T. Stuss & R. T. Knight (Eds.), *Principles of frontal lobe function* (pp. 311–325). New York: Oxford University Press.

Tzeng, O. J. L., & Cotton, B. (1980). A study-phase retrieval model of temporal coding. *Journal of Experimental Psychology: Human Learning and Memory, 6*, 705–716.

Wearden, J. H., & Culpin, V. (1998). Exploring scalar timing theory with human subjects. In V. De Keyser, G. d'Ydewalle, & A. Vandierendonck (Eds.), *Time and the dynamic control of behavior* (pp. 33–49). Seattle, WA: Hogrefe & Huber Publishers.

Wearden, J. H., & Ferrara, A. (1993). Subjective shortening in humans' memory for stimulus duration. *Quarterly Journal of Experimental Psychology: Comparative and Physiological Psychology, 46B*, 163–186.

Wheeler, M. A., Stuss, D. T., & Tulving, E. (1997). Toward a theory of episodic memory: The frontal lobes and autonoetic consciousness. *Psychological Bulletin, 121*, 331–354.

Zakay, D. (1990). The evasive art of subjective time measurement: Some methodological dilemmas. In R. A. Block (Ed.), *Cognitive models of psychological time* (pp. 59–84). Hillsdale, NJ: Erlbaum.

Zakay, D. (1992). The role of attention in children's time perception. *Journal of Experimental Child Psychology, 54*, 355–371.

Zakay, D. (1993). The roles of non-temporal information processing load and temporal expectations in children's prospective time estimation. *Acta Psychologica*, *84*, 271–280.
Zakay, D. (1998). Attention allocation policy influences prospective timing. *Psychonomic Bulletin and Review*, *5*, 114–118.
Zakay, D. (2000). Gating or switching? Gating is a better model of prospective timing (a response to 'switching or gating?' by Lejeune). *Behavioural Processes*, *50*, 1–7.
Zakay, D., & Block, R. A. (1996). The role of attention in time estimation processes. In M. A. Pastor & J. Artieda (Eds.), *Time, internal clocks and movement* (pp. 143–164). Amsterdam: North-Holland/ Elsevier Science.
Zakay, D., & Block, R. A. (1997). Temporal cognition. *Current Directions in Psychological Science*, *6*, 12–16.

Chapter 4: Sensory modalities and temporal processing*

Simon Grondin

Abstract

A central hypothesis in the field of time perception is that of a single clock. Judging time may depend on a central and unique timing device. At the heart of this issue is the question of the specificity of sensory modalities and their relation to time. The literature on timing and sensory modalities is reviewed, with an emphasis on duration discrimination. The review includes comparisons of the perceived duration of intervals marked by different sensory modes, and of the discrimination levels for different duration ranges. In addition to showing modality differences, it shows that perceived duration and discrimination levels also depend on the structure of intervals (filled versus empty; and the marker's length of empty intervals). In the case of empty intervals, not only does the discrimination level depend on the modality of markers, but it also depends on the fact that one (intramodal intervals) or two (intermodal intervals) modalities are involved in the task. It is argued that all of these effects on temporal processing can probably be accounted for by simple sensory (internal-marker hypothesis) or attentional explanations that are compatible with a single-clock hypothesis. Nevertheless, a fundamental question remains: Would it not be simpler to recognize the existence of some timing device with specific properties for each sensory modality, each being adapted to modality-specific environmental requirements (e.g., successive signals in the auditory mode, and processing of spatial information in the visual mode)?

Introduction

The purpose of this chapter is to present empirical facts and theoretical interpretations about temporal processing and sensory modalities. To that end, it will first explain what temporal processing actually means and provide a brief overview of the main issues addressed here.

Varieties of temporal processing

There is some confusion in experimental psychology about the use of the term *temporal processing*. Essentially, this term can refer to judgments about the order of arrival of sensory events (simultaneity, successiveness, order) or to judgments about

---

* This research is supported by a grant from the Natural Science and Engineering Research Council of Canada.

the temporal intervals defined by such events. It is judgments of the latter type that are discussed in this text. It should be noted, however, that the issue of modality is not only relevant to interval judgments, but of fundamental importance in the case of temporal order (Hirsh & Sherrick, 1961; Nicholls, 1996).

Even within the restricted scope emphasized in the present chapter, temporal processing can be approached from various perspectives: essentially, biologically versus cognitively oriented perspectives, based on animal versus human performances. This text adopts, for the most part, a cognitive/perceptual perspective, while reviewing work on human timing from a mainly psychophysical perspective. The mechanisms involved in the processing of time may depend on the range of duration under investigation (Grondin, Meilleur-Wells & Lachance, 1999). The processing of brief temporal intervals (in the vicinity of 1 s, or less) is the main, although not the only, concern here.

Varieties of issues

A good timing system is expected to provide two types of measure: a valid central tendency and low dispersion (reliability). In a series of trials, this system should generate values with a central tendency as close as possible to a real-time target, and the variability of these values should remain low (Grondin, 2001a). Sections 2 and 3 of this chapter are dedicated respectively to issues related to these two types of dependent variables, i.e., central tendency and variability.

The following analysis looks at what determines deviations from a time target (often referred to hereafter as *perceived duration*) or the variability of time estimates (mainly in duration discrimination tasks). Three independent variables are critical in this analysis. One is the structure of the intervals to be judged, which refers to the fact that an interval can be marked by a continuous signal of a given length (filled interval) or by the duration between two signals (empty interval). It is important to note that, in this case, "filled" does not mean the addition of a cognitive task during the interval, a strategy often used in the study of time perception with a cognitive emphasis. The second independent variable of particular interest here is the modality within which signals are delivered. As for the third independent variable, it is the method (single stimulus versus forced choice, for instance) adopted for presenting an interval or a series of intervals to be compared. This method is reported to determine the direction of results and conclusions regarding the effects exerted by the other two independent variables.

Central tendency

In this section, the dependent variable of interest is *perceived duration*. Indeed, it is not closeness to a real-time value *per se* that is the main focus here, but the comparison, for perceived duration, of intervals having different structures or being marked by signals delivered from different modalities.

Empirical facts

It is generally recognized that filled intervals are perceived as being longer than empty durations of the same length (Furukawa, 1979; Goldstone & Goldfarb, 1963). In fact, this inequality is sometimes reported to be huge. In a series of experiments, Craig (1973) asked participants to adjust the duration (empty interval) between two equal signals in order to make the duration equal to the first signal. For signal durations ranging from .1 to 1.2 s, the duration of the empty period had to be increased by a constant amount. The constant value was the same at all duration ranges, but differed slightly according to the signal's modality. In the tactile, auditory and visual modalities, it was 596, 657, and 436 ms, respectively.

It is usually reported that time intervals marked by auditory signals are judged as being longer than those marked by visual signals (Behar & Bevan, 1961; Goldstone & Lhamon, 1972; Goldstone & Goldfarb, 1964a, b). According to Lhamon and Goldstone (1974), this difference is observed more often under conditions where filled rather than empty intervals are used. Moreover, within the visual modality, it is the induction of movement that contributes the most to the increase in perceived duration, whereas in the auditory mode, sound intensity is the critical factor determining the perceived duration (Goldstone & Lhamon, 1974).

More recent research on modality differences has revealed that auditory-visual differences may depend on (1) the range of duration and (2) the intensity of signals. Walker and Scott (1981) used a reproduction method to show that, at 1 and 1.5 s, auditory signals lead to longer perceived duration than visual ones, and when both signals are presented, the duration produced is closer to that observed when only an auditory signal is presented. This is referred to by the authors as auditory *dominance*. However, at .5 s, the visual condition produces longer perceived duration if auditory signals are weak, and, if both signals are presented, visual dominance is observed (see also Goldstone, Boardman & Lhamon, 1959). Tanner, Patton and Atkinson (1965) also reported that visual intervals are perceived as being longer than auditory intervals of the same length. Recently, Wearden, Edwards, Fakhri and Percival (1998) reported longer perceived duration for audition than for vision with a wide range of durations (from about .1 to 1.2 s).

Empty intervals can be marked by signals delivered from different modalities. Such intervals are referred to as *intermodal* intervals. When intermodal intervals ranging from 250 to 750 ms are marked by a visual-auditory sequence, they are perceived as being shorter than those marked by an auditory-visual sequence (Grondin, Ivry, Franz, Perreault & Metthé, 1996; see also Hocherman & Ben-Dov, 1979). Interestingly, even if participants in experiments on intermodal intervals are asked to time the empty period (from the offset of Marker 1 to the onset of Marker 2), the duration of markers, either the first or the second one, has an influence on perceived duration. Intervals marked by 100-ms signals are perceived as being longer than those marked by 5-ms signals (see also Tsuzaki & Kato, 2000 and Woodrow, 1928 for other results on the effect of marker length on perceived duration). Moreover, in an experiment by Grondin and Rousseau (1991), intervals marked by a visual-auditory sequence were systematically perceived as being shorter than intramodal intervals (auditory-auditory or visual-visual sequences: Experiment 1). In the same experiment, the auditory-visual sequence was perceived as being longer than visual-auditory sequence, and auditory-visual sequences were perceived as being longer when the visual signal was presented at fovea rather than presented in periphery.

Intermodal intervals may also involve tactile intervals. Table 1 summarizes the results of another experiment by Grondin and Rousseau (1991: Experiment 2) regarding perceived duration. In this experiment, three marker-type intervals (one intramodal and two intermodal) were randomly presented within blocks of trials. In some blocks, the modality of the first marker was fixed (auditory, visual or tactile), while the modality of the second marker was randomized (auditory, visual or tactile); and in other blocks, the modality of the second marker was fixed (auditory, visual or tactile), while the modality of the first marker was randomized (auditory, visual or tactile). Among the conclusions drawn from this experiment, it was reported that the intermodal interval tended to be long when the first marker was tactile, and short when the first marker was visual. However, and most importantly, it was impossible to draw any firm conclusions about the effect of a modality as a first or second marker when both intramodal and intermodal intervals were taken into account.

First marker is fixed; second marker is randomized		
AA = .4903	VA = .4534	TA = .6814
AV = .5621	VV = .5279	TV = .6316
AT = .5254	VT = .4307	TT = .4503
M = .5269	*M* = .4707	*M* = .5878
First marker is randomized; second is marker fixed		
AA = .4810	AV = .4641	AT = .5880
VA = .3278	VV = .4707	VT = .4255
TA = .6330	TV = .5982	TT = .4844
M = .4806	*M* = .5110	*M* = .4993

Note- A = auditory, V = visual, T = tactile, *M* = mean.

Table 1: Mean probability of responding "long" under each marker-type condition. Adapted from Grondin and Rousseau (1991; Experiment 2)

The perceived duration of intermodal empty intervals can be altered by adding a brief sensory signal. If a tactile signal is added, the interval is perceived as being longer if it is delivered closer to the second marker than to the first one (Grondin & Rousseau, 1999; see also Grondin, 2001b). However, if a brief visual signal is added during an empty interval, it is the signal's occurrence *per se* rather than its location that will exert the most influence on perceived duration: adding a brief visual signal increases perceived duration.

Lastly, the method used to present intervals is critical. Perceived duration can be severely distorted by using sequences of intervals. These distortions occur in both the visual modality (Arao, Suetomi, & Nakajima, 2000; Rose & Summers, 1995) and the auditory modality (Nakajima, ten Hoopen, Hilkhuysen & Sasaki, 1992; Nakajima, ten Hoopen & van der Wilk, 1991; ten Hoopen, Hartsuiker, Sasaki, Nakajima, Tanaka & Tsumura, 1995; ten Hoopen, Hilkhuysen, Vis, Nakajima, Yamauchi & Sasaki, 1993). Indeed, only a slight variation of a preceding interval changes the perceived duration of an interval to be discriminated, either marked by auditory or visual signals (Grondin & Rammsayer, in press).

Theoretical explanations

How can such structure or modality effects on perceived duration be explained? The literature on time perception offers two main types of explanation. One is based on the properties of a so-called internal clock, often referred to as a pacemaker–counter device (see Killeen & Taylor, 2000; or Killeen & Weiss, 1987). The pacemaker is said to emit pulses that are accumulated by the counter. This accumulation constitutes the basis of subjective time: the more pulses accumulated, the longer the perceived duration. According to Wearden et al. (1998), sound and light have different effects on the pacemaker's rate of emission. Sound provokes a faster rate of emission than light does, with the result that there is a potentially greater accumulation of pulses by the counter with auditory signals than with visual ones. Modality-based differences in perceived duration are also reported to depend partly on the involvement of memory processes (see Allan, 1998; Penney, Gibbon & Meck, 2000).

Another explanation, consistent with the hypothesis of an internal clock such as the one described above, locates variations in perceived duration at the level of marking, i.e., the way in which intervals are marked internally (internal-marker hypothesis, see Grondin, 1993). As indicated in Figure 1 (left panel), filled intervals may be perceived as being longer than empty ones because of variations that occur in the internal timekeeping period at the beginning and at the end of the process. The filled/-empty effect reported by Craig (1973) and described above can be explained, at least in part, by a sensory cause of this type. A delay in ending the timekeeping period of the first (filled) interval would also cause a delay in starting the timekeeping period of the following (empty) interval. Both of these effects would concur to make a filled interval seem longer than an empty one.

In the case of intermodal intervals, certain specific characteristics of modalities could lead to variations in perceived duration (see Figure 1, right panel). This simple sensory explanation could account for a result such as longer perceived duration with an auditory-visual sequence than with a visual-auditory sequence. The beginning of the internal timekeeping period may occur earlier with an auditory signal than it does with a visual one. Moreover, the end of the timekeeping activity would occur earlier if Marker 2 were auditory instead of visual: this hypothesis is plausible considering that reaction time is faster to auditory signals than to visual ones. However, this general sensory explanation would not hold for some of the results described earlier, involving both intramodal and intermodal intervals or differing marker-length conditions.

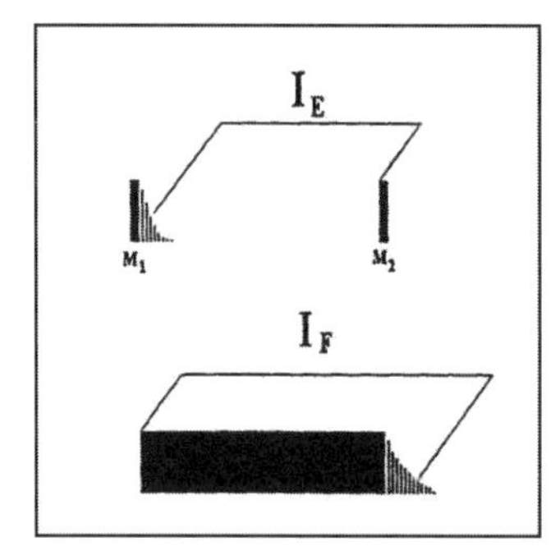

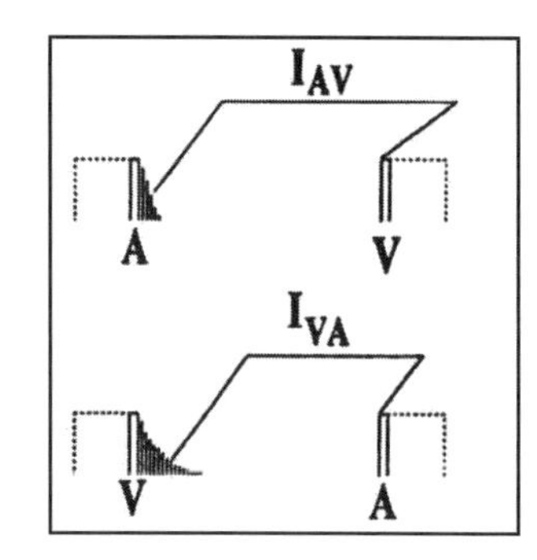

Figure 1: Illustration of an internal-marker hypothesis applied to the filled/empty issue (left panel) and to an intermodal interval issue (right panel). The physical duration from the physical offset of Marker 1 and the physical onset of Marker 2 of the empty interval and, the physical duration from the physical onset to the physical offset of the filled interval, are equal. However, the internal timekeeping period is argued to differ. As well, the physical duration between the AV and VA sequences is the same, but not the perceived duration (I = Interval; M = Marker; E = Empty; F = Filled; A = Auditory; V = Visual).

Variability

The dispersion of a series of judgments (variability) is expressed in various ways in the time perception literature. Viewed from a psychophysical perspective, this dependent variable is most often associated with the concept of threshold. In duration discrimination tasks, degrees of variability (performance-sensitivity) are expressed as a percentage of correct responses, d' or as threshold estimates (i.e., Weber fractions) depending on the method used.

Empirical facts

Both the modalities used to mark time and the issue of filled versus empty structures exert a major influence on time interval discrimination capabilities. Both issues were addressed in an experiment conducted by the author ten years ago, of which the main results are reported in Figure 2 (based on Grondin, 1993, Table 7). As indicated in this figure, discrimination is better (lower Weber fractions) with empty intervals than with filled ones when intervals are brief. The figure also indicates that discrimination is much better with auditory than with visual intervals. Finally, Weber fractions are almost constant in all conditions with longer intervals (note that the participants in the above-mentioned experiment were not asked to refrain from using explicit counting of numbers).

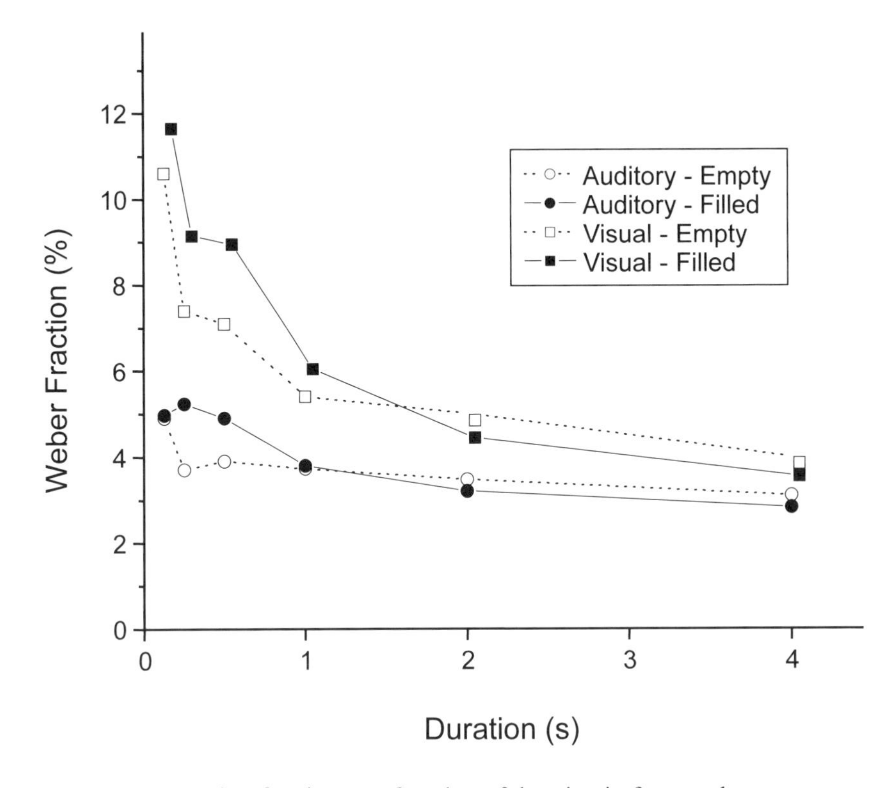

Figure 2: Weber fraction as a function of duration in four marker-type conditions (Data from Grondin, 1993, Table 7)

However, the foregoing conclusion regarding the filled/empty issue is not that simple considering that other studies have revealed better discrimination with filled intervals than with empty ones. Table 2 summarizes the many studies on this question found in the literature. The most commonly replicated finding on the superiority of filled intervals over empty ones comes from the laboratory of T. Rammsayer, who studied mainly 50-ms intervals using an adaptive procedure (Rammsayer & Lima, 1991; Rammsayer & Skrandies, 1998; Skrandies & Rammsayer, 1995).

One critical determinant of the quality of the discrimination of empty intervals is the length of markers. If markers are long, discrimination is impaired and filled intervals are discriminated much better than empty ones are (Kato & Tsuzaki, 1994). Indeed, if markers are too long, it might not be an interval that is discriminated, but a gap. The most complete report on this question is probably the one by Rammsayer and Leutner (1996), whose results are summarized in Table 3. Many other reports show that the perception of time, or rhythm, is influenced by a marker's duration (Divenyi & Sachs, 1978; Handel, 1993; Penner, 1976; Yamashita & Nakajima, 1999).

Author(s) (year)	Duration (ms)	Method (FC: Forced choice)	Markers' length (Empty interval)	Difference Threshold (ms) or % correct Empty	 Filled	 Gap
AUDITORY MODE						
Fitzgibbons & Gordon-Salant (1994)	St = 250	Adaptive (FC)	250 – 250	59	57	
Grondin (1993)	241 vs 259	Single stimulus	20 – 20	81%	70%	
	50 vs 62	Single stimulus	20 – 20	76%	78%	
	241 vs 259	FC	20 – 20	75%	72%	
		Single stimulus	20 – 20	72%	66%	
	St = 250	Adaptive (FC)	20 – 20	36	33	
	St = 50	Adaptive (FC)	20 – 20	11	11	
Grondin, Meilleur-Wells, Ouellette & Macar (1998)	350 vs 400	Single stimulus	20 – 20	85%	78%	
	St = 400	Adaptive (FC)	20 – 20	39	46	
	700 vs 800	Single stimulus	20 – 20	82%	77%	
	St = 800	Adaptive (FC)	20 – 20	80	81	
Kato & Tsuzaki (1994)	St = 170	Constant (FC)	1400 – 1100 (Flankers)		24 – 63	
			1400 – 1500 – 1100 (Gap)			75
			21 – 21	19	17 – 19	
Rammsayer & Lima (1991)	St = 50	Adaptive (FC)	3 – 3	20	7	
		+ cognition load		19	7	
		+ sound backward masking		46	21	
Rammsayer & Skrandies (1997)	St = 50	Adaptive (FC)	3 – 3	11	6	30
			1500 – 900 – 1000 (Gap)			
Skrandies & Rammsayer (1995)	St = 50	Adaptive (FC)	1500 – 900 – 1000 (Gap)		14	68

Table 2 /…continued next page

Author(s) (year)	Duration (ms)	Method (FC: Forced choice)	Markers' length (Empty interval)	Difference Threshold (ms) or % correct Empty	Filled	Gap
VISUAL MODE						
Grondin (1993)	225 vs 275	Single stimulus	20 – 20	82%	77%	
	50 vs 80	Single stimulus	20 – 20	84%	76%	
	St = 250	Adaptive (FC)	20 – 20	57	87	
	St = 50	Adaptive (FC)	20 – 20	21	26	
Grondin, Meilleur-Wells, Ouellette & Macar (1998)	350 vs 400	Single stimulus	20 – 20	69%	62%	
	St = 400	Adaptive (FC)	20 – 20	63	84	
	700 vs 800	Single stimulus	20 – 20	70%	67%	
	St = 800	Adaptive (FC)	20 – 20	118	109	

Table 2: A summary of results for duration discrimination experiments involving direct comparison of empty versus filled intervals.

Experiment	Standard (ms)	Marker 1 (ms)	Marker 2 (ms)	I S I (ms)	Difference threshold (ms)
1	50	3	3	894	17
	50	30	30	840	18
	50	300	300	300	34
	50	1-s signal – gap – 1-s signal – gap – 1-s signal			39
2	50	3	3	900	18
	50	30	30	900	17
	50	300	300	900	35
	50	1-s signal – gap – .9-s signal – gap – 1-s signal			37
3	50	75	75	900	19
	50	150	150	900	21
	50	225	225	900	32
4	50	300	3	900	66
	50	3	300	900	65
	300	3	3	900	45
	300	300	300	900	85

Table 3: Effects of markers' length on difference threshold of short auditory empty time intervals. Adapted from Rammsayer and Leutner (1996).

Another critical factor in the filled versus empty issue is the method used to estimate discrimination capabilities. The composition of intervals, i.e., whether they consist of one (filled) or two (empty) signals, is of fundamental importance here. In experiments involving a forced choice procedure, only two events are presented to participants if filled intervals are used, while four events are required if two empty intervals have to be discriminated. In the latter case, the presence of a third empty interval, i.e., the inter-stimulus interval, may make the discrimination process more confusing. This method effect has been shown to exert a clear influence on results relating to the filled/empty issue (Grondin, Meilleur-Wells, Ouellette & Macar, 1998). However, the influence also depends on the range of duration: more confusion is created by a forced choice situation with .8-s intervals than with .4-s ones. With a single-stimulus mode of presentation, superior discrimination of empty intervals is clearly observed at both .4 and .8 s.

Some of the above-mentioned studies on the filled versus empty issue also revealed much better discrimination of auditory intervals over visual ones (Grondin, 1993; Grondin et al., 1998). Other reports are also consistent with these results (Grondin, Ouellet & Roussel, 2001; Rousseau, Poirier & Lemyre, 1983). However, less information is available regarding the discrimination of tactile intervals. Goodfellow (1934) reported that tactile markers provide slightly better results than visual markers do for empty intervals lasting about 1 s, but others have reached the opposite conclusion for intervals lasting 250 ms (Grondin & Rousseau, 1991, Experiment 2). Figure 3 shows results based on other data for those two ranges of duration (Grondin, 1994). In the experiments illustrated in Figure 3, participants were asked to discriminate 225 ms from 275 ms; and 925 ms from 1075 ms. Empty intervals were presented according to a single-stimulus method. The results revealed that, for both ranges of duration, discrimination is much better in the auditory condition. These results are consistent with the above-mentioned reports on tactile versus visual markers.

Using intermodal intervals clearly impairs discrimination performances. It is well established that, for brief intervals (< 1 s), combinations involving both auditory and visual signals lead to poorer performances than those involving two visual signals (Grondin & Metthé, 1993; Grondin & Rousseau, 1991; Rousseau et al., 1983). Figure 3 also shows that the same conclusion applies, at 250 ms, to combinations involving a tactile marker.

However, using two tactile stimuli at 1 s does not lead to better discrimination than the use of one tactile and one auditory signal, although it does generate better performances than using sequences involving a tactile and a visual signal.

Empirical reports on intermodal discrimination are restricted to a range of brief durations. Nevertheless, one can find data on intermodal performances for intervals ranging from .125 to 4 s. For example, Figure 4 of a report on a previous experiment conducted by the author (Grondin, 1996) shows data derived from threshold estimates based on multi-point psychometric functions. It reveals the very high Weber fractions obtained with very brief intervals for both visual-auditory and auditory-visual sequences. The Weber fractions were even higher for one of the participants in the experiment, whose results are not shown in the figure, and constant for all participants when longer intervals were used (participants were not asked to refrain from using explicit counting of numbers).

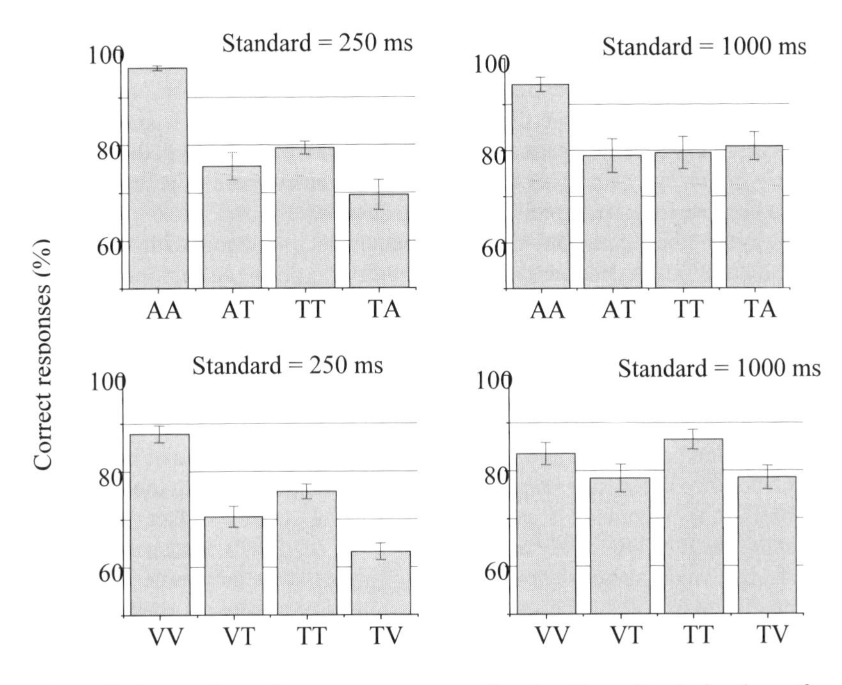

Figure 3: Proportion of correct responses for duration discrimination of empty intervals in several marker-type conditions. Discrimination at 250 ms is between 225 and 275 ms; and at 1 s, is between 925 and 1075 ms (A=Auditory; V = Visual; T = Tactile). Data from Grondin (1994).

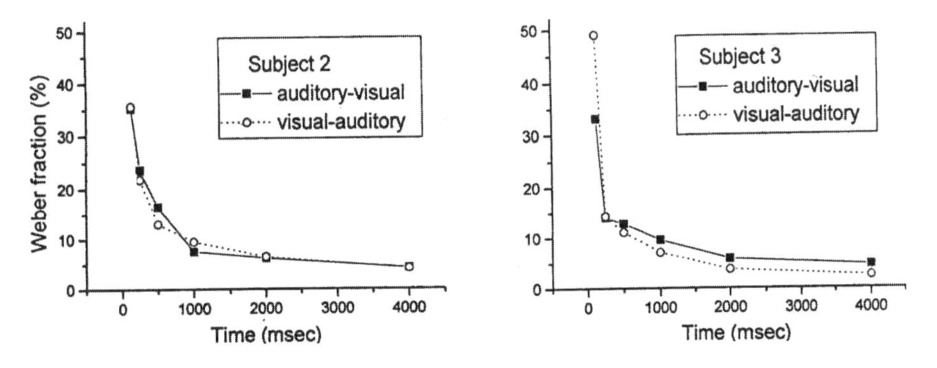

Figure 4: Weber fraction as a function of duration for two participants in two intermodal conditions. Data from Grondin (1996)

Theoretical explanations

Different explanations for differences in discrimination performances (variability) with filled versus empty intervals lead to different predictions. One explanation is based on the internal-marker hypothesis described above (see Figure 1). If variability is proportional to the period of accumulation of pulses (more variability with longer intervals), as should be the case with a pacemaker-counter perspective, there should be more variability with filled intervals if the marking hypothesis is correct. In other words, empty intervals should be discriminated more effectively than filled intervals because they are perceived as being shorter.

However, other explanations predict better discrimination with filled intervals than with empty ones. Rammsayer and Lima (1991) proposed that the continuity of signal stimulation provided by filled intervals should increase the rate of pulse emission by the pacemaker. The possibility of accumulating more pulses within a given period would result in fine temporal resolution. The finer resolution obtained with filled than with empty intervals is one of the explanations that underpins predictions regarding better discrimination with filled intervals. Another theoretical reason is the above-mentioned confusion that might be created by the sequence of brief events involved when two successive empty intervals are presented. Indeed, the critical role of attention in time judgments is well known (Grondin, 2001b). In fact, it is the potential displacement of attention caused by the series of signals associated with empty intervals that would make the accumulation of pulses in the counter less reliable.

The differences observed under various modality conditions (better discrimination in audition) when filled intervals or empty intramodal intervals are used can be explained reasonably well by differences in the clarity/efficiency of the signals themselves for marking the beginning or end of timekeeping periods. However, such a sensory-based explanation cannot alone account for the impairment caused by the use of intermodal conditions. This sensory explanation would predict that using the most efficient signal (auditory) as Marker 1 or Marker 2 would lead, in the case of an auditory-visual or visual-auditory sequence for example, to a level of discrimination that was situated somewhere between the levels obtained with auditory-auditory or visual-visual conditions. However, this prediction is incorrect. Additional explanations are required to account for the intramodal versus intermodal difference. One view posits that intramodal sequences are advantaged by the similarity of both signals, the efficiency of the second signal being enhanced by the preceding arrival of the first one (prior entry explanation). An alternative view is that intramodal intervals are not advantaged in any way, but that intermodal sequences loose efficiency because of the need to switch attention between modalities.

Interestingly, all of these explanations are consistent with the single-clock hypothesis. However, some data, such as those on relative perceived durations with multi-modal conditions, reported briefly above, are difficult to explain with this theory (see Grondin & Rousseau, 1991). This observation, along with several other reasons given below, might argue for the adoption of different points of view on the nature of human timing mechanisms.

One alternative: A modality-specific approach

Although a single-clock perspective is viable, other considerations suggest that alternative views on human timing should be explored. The single-clock hypothesis (pacemaker-counter device) requires a large variety of post hoc explanations regarding modality, marker or sequence effects on perceived duration or discrimination levels, as well as the involvement of memory mechanisms and attention resources and properties. Moreover, this clock perspective predicts that Weber's law will hold, i.e., that proportional growth of variability will be observed as a function of time (also referred to as the scalar property). Several violations of this law are found in the time perception literature (Grondin, 2001a), especially at points between intervals of 1 to 1.5 s (see Grondin, in press-b).

Furthermore, some timing researchers propose that more than one timing mechanism may be involved. For instance, Jones and Boltz (1989) do not adopt a central-timer perspective, but emphasize an ecological perspective where internal capabilities for attending to external signals help to capture environmental regularities in sensory events. It is these regularities that generate plausible expectations regarding future events in time. Other examples of different perspectives on timing include those of Rammsayer (this volume, chapter 6), who highlights the fundamental difference (cognitively mediated or not) observed between the mechanisms involved in processing 50 and 1000-ms intervals (Rammsayer, Hennig, Haag & Lange, 2001; Rammsayer & Lima, 1991); and Staddon, who locates all variance in the timing process at the memory level (Staddon & Higa, 1996, 1999).

An alternative way to approach the timing issue from the perspective of sensory modalities is to look at what characterizes each modality. In the auditory mode, temporal resolution is remarkably good, which is not surprising when we consider that successive events, in speech or music for example, need to be processed rapidly. In this context, it is imperative to study interval discrimination in the context of tempo or rhythm. For instance, it is known that presenting more standard intervals enhances discrimination performance in the auditory modality (Drake & Botte, 1993), although it does not necessarily do so in the visual modality if intervals are very short (in the range of 300 ms instead of 900 ms, for instance, see Grondin, 2001c).

Discrimination of auditory intervals is even better if they are embedded in a musical context. For example, if only the time between the notes of a Bach prelude is changed, the Weber fraction for this change discrimination is below 1% for musically trained participants (Grondin & Laforest, 2002).

The fact that space is taken into account in the visual modality is no doubt of critical importance (see Grondin, in press-a). Most timing research in the perspective of an internal clock uses visual stimuli delivered from a single source. If more than one visual source is used, several questions may emerge. For instance, three potential visual signals could be located on the same vertical plane in front of a participant. If space is taken into account in judging time, the visual perception cues may interfere with temporal judgments. In one experiment, three LEDs were located at fovea (F), just above (A) or just below (B). The LEDs marked randomly presented brief empty intervals, forming F-A, B-F or B-A sequences (or A-F, F-B or A-B sequences in another series of trials). In this experiment, it was not the height of the markers in the visual field that influenced perceived duration the most, but the total space between markers (Guay & Grondin, 2001). Surprisingly, contrary to what is often reported for the kappa effect, shorter perceived duration was observed with the A-B or B-A sequences.

However, if visual sources are located in different visual fields, i.e., left versus right, the question of cerebral hemispheric specialization or dominance for temporal processing arises. It is known that discrimination of empty intervals is better if they are marked by signals delivered at the same location, either to the right or to the left, instead of on different sides (Grondin, 1998). However, when both signals are delivered on the same side (left or right), there do not seem to be any differences in discrimination performances, for either filled or empty intervals, that may be attributed to hemispheric differences (Grondin, Lapointe, Guay & Ouellet, 2002). However, as far as perceived duration is concerned, left-right sequences are perceived as being longer than right-left ones (Grondin, 1998).

Many questions remain unanswered with regard to time-sense relations. Although the single-clock perspective has proved useful over the past 30 years of research on timing and time perception mechanisms, more effort should now be devoted to elaborating an alternative perspective where explanations would be based on the nature of events in the environment and on potential cognitive adjustments. More effective integration of the main principles derived from the perception field into time research could play a key role in acquiring a new understanding of psychological time.

References

Allan, L. G. (1998). The influence of the scalar timing model on human timing research. *Behavioural Processes*, *44*, 101–117.

Arao, H., Suetomi, D., & Nakajima, Y. (2000). Does time-shrinking take place in visual temporal patterns? *Perception*, *29*, 819–830.

Behar, I., & Bevan, W. (1961). The perceived duration of auditory and visual intervals: Cross-modal comparison and interaction. *American Journal of Psychology*, *74*, 17–26.

Craig, J. C. (1973). A constant error in the perception of brief temporal intervals. *Perception and Psychophysics*, *13*, 99–104.

Divenyi, P. L., & Sachs, R. M. (1978). Discrimination of time intervals bounded by tone bursts. *Perception and Psychophysics*, *24*, 429–436.

Drake, C., & Botte, M-C. (1993). Tempo sensitivity in auditory sequences : Evidence for a multiple-look model. *Perception and Psychophysics*, *54*, 277–286.

Fitzgibbons, P. J., & Gordon-Salant, S. (1994). Age effects on measures of auditory duration discrimination. *Journal of Speech and Hearing Research*, *37*, 662–670.

Furukawa, M. (1979). A study on the difference in the visual and auditory temporal judgment (II). *Tohoku Psychologica Folia*, *38*, 18-28.

Goldstone, S., Boardman, W. K., & Lhamon, W.T. (1959). Intersensory comparisons of temporal judgments. *Journal of Experimental Psychology*, *57*, 243–248.

Goldstone, S., & Goldfarb, J. L. (1963). Judgment of filled and unfilled durations: Intersensory factors. *Perceptual and Motor Skills*, *17*, 763–774.

Goldstone, S., & Goldfarb, J. L. (1964a). Auditory and visual time judgment. *Journal of General Psychology*, *70*, 369–387.
Goldstone, S., & Goldfarb, J. L. (1964b). Direct comparison of auditory and visual durations. *Journal of Experimental Psychology*, *67*, 483–485.
Goldstone, S., & Lhamon, W. T. (1972). Auditory-visual differences in human temporal judgment. *Perceptual and Motor Skills*, *34*, 623–633.
Goldstone, S., & Lhamon, W. T. (1974). Studies of auditory-visual differences in human time judgment: 1. Sounds are judged longer than lights. *Perceptual and Motor Skills*, *39*, 63–82.
Goodfellow, L. D. (1934). An empirical comparison of audition, vision, and touch in the discrimination of short intervals of time. *American Journal of Psychology*, *46*, 243–258.
Grondin, S. (1993). Duration discrimination of empty and filled intervals marked by auditory and visual signals. *Perception and Psychophysics*, *54*, 383–394.
Grondin, S. (1994). About the influence of the sensory modes on duration discrimination. In L. Ward (Ed.), *Fechner Day 94: Proceedings of the 10th Annual Meeting of the International Society for Psychophysics* (pp. 42–47). Vancouver : International Society for Psychophysics.
Grondin, S. (1996). Discriminating intermodal time intervals. In S. Masin (Ed.), *Fechner Day 96: Proceedings of the 12th Annual Meeting of the International Society for Psychophysics* (pp. 275–280). Padua: International Society for Psychophysics.
Grondin, S. (1998). Judgments of the duration of visually marked empty intervals: Linking perceived duration and sensitivity. *Perception and Psychophysics*, *60*, 319–330.
Grondin, S. (2001a). From physical time to the first and second moments of psychological time. *Psychological Bulletin*, *127*, 22–44.
Grondin, S. (2001b). Time psychophysics and attention. *Psychologica*, *28*, 177–191.
Grondin, S. (2001c). Discriminating time intervals presented in sequences marked by visual signals. *Perception and Psychophysics*, *63*, 1214–1228.
Grondin, S. (in press). Processing time between visual events. *Brazilian Journal of Ophthalmology*.
Grondin, S. (in press). Studying psychological time with Weber's law. In R. Buccheri, M. Saniga & M. Stuckey (Eds.), *The Nature of time: Geometry, physics and perception*. Dordrecht, The Netherlands: Kluwer Academic.
Grondin, S., Ivry, R. B., Franz, E., Perreault, L., & Metthé, L. (1996). Markers' influence on the duration discrimination of intermodal intervals. *Perception and Psychophysics*, *58*, 424–433.
Grondin, S., & Laforest, M. (2002). Discriminating slow tempo variations in a musical context. In J. A. Da Silva, E. H. Matsushima & N. P. Ribeiro Filho (Eds.), *Fechner Day 2002: Proceedings of the 18th Annual Meeting of the International Society for Psychophysics* (pp. 398–403). Rio de Janeiro, Brazil: International Society for Psychophysics.
Grondin, S., Lapointe, M., Guay, I., & Ouellet, B. (2002). About hemispheric asymmetries in relation to the discrimination of visually marked intervals [Abstract]. *Brain and Cognition*, *48*, 248.
Grondin, S., Meilleur-Wells, G., & Lachance, R. (1999). When to start explicit counting in time-intervals discrimination task: A critical point in the timing process of humans. *Journal of Experimental Psychology: Human Perception and Performance*, *25*, 993–1004.
Grondin, S., Meilleur-Wells, G., Ouellette, C., & Macar, F. (1998). Sensory effects on judgments of short-time intervals. *Psychological Research*, *61*, 261–268.

Grondin, S., & Metthé, L. (1993). Procedural effects on duration discrimination. In A. Garriga-Trillo, P. Minon, C. Garcia-Gallego, P. Lubin, J. Morino, & A. Villarino (Eds.). *Fechner Day 93: Proceedings of the 9th Annual Meeting of the International Society for Psychophysics* (pp.107–112). Palma de Mallorca, Spain: International Society for Psychophysics.

Grondin, S., Ouellet, B., & Roussel, M.-E. (2001). About optimal timing and stability of Weber fraction for duration discrimination. *Acoustical Science & Technology*, *22*, 370–372.

Grondin, S., & Rammsayer, T. H. (in press). Variable foreperiods and duration discrimination. *Quarterly Journal of Experimental Psychology*.

Grondin, S., & Rousseau, R. (1991). Judging the relative duration of multimodal short empty time intervals. *Perception and Psychophysics*, *49*, 245–256.

Grondin, S., & Rousseau, R. (1999). A modality specific filled-duration effect. In P. R. Killeen & W. Uttal (Eds.), *Fechner Day 99: Proceedings of the 15th Annual Meeting of the International Society for Psychophysics* (pp. 256–261). Tempe, U.S.A.: The International Society for Psychophysics.

Guay, I., & Grondin, S. (2001). Influence of time interval categorization of distance between markers located on a vertical plane. In E. Sommerfeld, R. Kompass, & T. Lachmann (Eds.), *Proceedings of the 17th Annual Meeting of the International Society for Psychophysics* (pp. 391–396). Berlin: Pabst Science Publishers.

Handel, S. (1993). The effect of tempo and tone duration on rhythm discrimination. *Perception and Psychophysics*, *54*, 370–382.

Hirsh, I. J., & Sherrick, C. E. (1961). Perceived order in different sense modalities. *Journal of Experimental Psychology*, *62*, 423–432.

Hocherman, S., & Ben-Dov, G. (1979). Modality-specific effects on discrimination of short empty time intervals. *Perceptual and Motor Skills*, *48*, 807–814.

Jones, M. R., & Boltz, M. G. (1989). Dynamic attending and responses to time. *Psychological Review*, *96*, 459–491.

Kato, H., & Tsuzaki, M. (1994). Intensity effect on discrimination of auditory duration flanked by preceding and succeeding tones. *Journal of Acoustical Society of Japan (E)*, *15*, 349–351.

Killeen, P. R., & Taylor, T. (2000). How the propagation of error through stochastic counters affects time discrimination and other psychophysical judgments. *Psychological Review*, *107*, 430–459.

Killeen, P. R., & Weiss, N. A. (1987). Optimal timing and the Weber function. *Psychological Review*, *94*, 455–468.

Lhamon, W. T., & Goldstone, S. (1974). Studies on auditory-visual differences in human time judgment: 2. More transmitted information with sounds than lights. *Perceptual and Motor Skills*, *39*, 295–307.

Nakajima, Y., ten Hoopen, G., Hilkhuysen, G., & Sasaki, T. (1992). Time-shrinking: A discontinuity in the perception of auditory temporal patterns. *Perception and Psychophysics*, *51*, 504–507.

Nakajima, Y., ten Hoopen, G., & van der Wilk, R. G. H. (1991). A new illusion of time perception. *Music Perception*, *8*, 431–448.

Nicholls, M. E. R. (1996). Temporal processing asymmetries between the hemispheres: Evidence and implications. *Laterality*, *1*, 97–137.

Penner, M. (1976). The effect of marker variability on the discrimination of temporal intervals. *Perception and Psychophysics*, *19*, 466–469.

Penney, T. B., Gibbon, J., & Meck, W. H. (2000). Differential effects of auditory and visual signals on clock speed and temporal memory. *Journal of Experimental Psychology: Human Perception and Performance, 26*, 1770–1787.

Rammsayer, T. H., Hennig, J., Haag, A., & Lange, N. (2001). Effects of noradrenergic activity on temporal information processing in humans. *Quarterly Journal of the Experimental Psychology, 54B*, 247–258.

Rammsayer. T. H., & Lima, S. D. (1991). Duration discrimination of filled and empty auditory intervals: Cognitive and perceptual factors. *Perception and Psychophysics, 50*, 565–574.

Rammsayer, T. H., & Leutner, D. (1996). Temporal discrimination as a function of marker duration. *Perception and Psychophysics, 58*, 1213–1223.

Rammsayer, T. H., & Skrandies, W. (1998). Stimulus characteristics and temporal information processing: Psychophysical and electrophysiological data. *Journal of Psychophysiology, 12*, 1–12.

Rose, D., & Summers, J. (1995). Duration illusions in a train of visual stimuli. *Perception, 24*, 1177–1187.

Rousseau, R., Poirier, J., & Lemyre, L. (1983). Duration discrimination of empty time intervals marked by intermodal pulses. *Perception and Psychophysics, 34*, 541–548.

Skrandies, W., & Rammsayer, T. H. (1995). The perception of temporal structure and auditory evoked brain potential. *Biological Psychology, 40*, 267–280.

Staddon, J. E. R., & Higa, J. J. (1996). Multiple time scales in simple habituation. *Psychological Review, 103*, 720–733.

Staddon, J. E. R., & Higa, J. J. (1999). Time and memory: Towards a pacemaker-free theory of interval timing. *Journal of the Experimental Analysis of Behavior, 71*, 215–251.

Tanner, T. A., Patton, R. M., Atkinson, R. C. (1965). Intermodality judgments of signal duration. *Psychonomic Science, 2*, 271–272.

ten Hoopen, G., Hartsuiker, R., Sasaki, T., Nakajima, Y., Tanaka, M., & Tsumura, T. (1995). Auditory isochrony: Time shrinking and temporal patterns. *Perception, 24*, 577–593.

ten Hoopen, G., Hilkhuysen, G., Vis, G., Nakajima, Y., Yamauchi, F., & Sasaki, T. (1993). A new illusion of time perception–II. *Music Perception, 11*, 15–38.

Tsuzaki, M., & Kato, H. (2000). Shrinkage of perceived tonal duration produced by extra sounds: Effects of spectral density, temporal position, and transition direction. *Perception, 29*, 989–1004.

Walker, J. T., & Scott, K. J. (1981). Auditory-visual conflicts in the perceived duration of lights, tones, and gaps. *Journal of Experimental Psychology: Human Perception and Performance, 7*, 1327–1339.

Wearden, J. H., Edwards, H., Fakhri, M., & Percival, A. (1998). Why "sounds are judged longer than lights": Application of a model of the internal clock in humans. *Quarterly Journal of Experimental Psychology, 51B*, 97–120.

Woodrow, H. (1928). Behavior with respect to short temporal stimulus forms. *Journal of Experimental Psychology, 11*, 167–198.

Yamashita, M., & Nakajima, Y. (1999). The effect of marker duration on time-shrinking. In S. Won Yi (Ed.), *Music, Mind, and Science* (pp. 211–218). Seoul: National University Press.

Chapter 5: Notable results regarding temporal memory and modality

FLORIAN KLAPPROTH

Abstract

The chapter reports three experiments using a temporal generalization paradigm. Two experiments (Experiment 1 and 2) examined whether stimulus modality (auditory versus visual) affects the retrieval of subjective duration from memory. Participants had to decide whether the previously-learned standard duration (400 ms) occurred in the context of comparison stimuli. Two major results were found. When subjects initially learn time-intervals and are then required to remember them subsequently, discrimination is more accurate if the training and testing stimuli are in the same modality than if they are of different modalities. Furthermore, if both the modality of training and the modality of testing are different, subjects systematically underestimate the test durations, i.e., temporal generalization gradients (the proportion of identifications of a stimulus as the standard, plotted against stimulus duration) shift to the right. Experiment 3 examined whether variation of intra-modal stimulus quality (sinus tone versus white noise) also affected performance in the testing phase. As expected, changing the quality between the training phase and the testing phase biased discrimination, but the effect of underestimation did not occur. Therefore, underestimation of presented duration seems to depend on the change of modality.

The effect of underestimation points to at least two conclusions. First, it is likely that the memory representation of the standard duration is associated with the modality in which it was acquired. Second, the observed shift could be interpreted as a result of a delayed timing process.

Introduction

Investigations of the dependence of duration discrimination on modality are common and the results are well known. For example, it has been frequently reported that (1) visual stimuli are usually judged to be shorter than auditory stimuli of physically equal duration (e.g., Goldstone & Goldfarb, 1964; Goldstone & Lhamon, 1974; Sebel & Wilsoncroft, 1983) and (2) that visual intervals are judged with less accuracy than auditory intervals (e.g., Collier & Logan, 2000; Glenberg, Mann, Altman, Forman, & Procise, 1989; Rousseau & Rousseau, 1996; Wearden, Edwards, Fakhri, & Percival, 1998).

Recently-developed explanations of the modality effect are often embedded within the framework of *scalar timing theory* (Gibbon, 1991; Gibbon & Church, 1984). *Scalar timing theory* distinguishes three stages of the processing of time intervals: an internal clock stage, incorporating a pacemaker, a switch, and an accumulator; a memory stage; and a decision stage. At the internal clock stage physical time is encoded proportionally into a quantity of pulses. These pulses are transferred into an accumulator via a switch that closes when a to-be-timed signal is present. The longer a stimulus is in time, the more pulses are generated and accumulated.

The finding that visual stimuli are judged shorter than auditory ones can be attributed to clock-speed differences between vision and audition (e.g., Penney, Allan, Meck, & Gibbon, 1998; Wearden et al., 1998). The accumulated value of perceived time will be smaller when the timed signal is visual than when it is auditory because the rate of producing pulses is smaller when the modality of the encoded duration is visual rather than auditory. Thus, if auditory and visual accumulated values are compared, the visual signal will seem shorter than the auditory one.

Differences in accuracy between vision and audition can be explained by a differential variability in the operation of the switch in the pacemaker-accumulator system (cf. Wearden, 1999; Wearden et al., 1998; see also Wearden, this volume, chapter 2). However, relatively little interest has been shown in the question of whether duration values of auditory and visual stimuli are represented differentially in human memory (exceptions are the Hocherman & Ben-Dov study, 1979, and the studies from Penney et al., 1998, 2000). Hence, the topic of the first experiment was to examine whether stimulus modality affects the retrieval of subjective duration from long-term memory. In this context, long-term memory refers to a memory that stores values of duration for the period of the whole task that is used.

Experiment 1

In the *temporal generalization* task used participants had to identify a previous-learned standard duration, which is initially learned, then presented intermixed with comparison stimuli of both shorter and longer durations.

The task was divided into a training phase and a testing phase. The modality of the two experimental phases was varied between participants, and duration of presented stimuli was varied within participants. Thus, four conditions were used: AUD-AUD, VIS-VIS, AUD-VIS, and VIS-AUD.

Hypotheses

First, analogous to the finding that sounds are judged with greater accuracy than visual stimuli, we examined whether performance in the testing phase would be better when the time interval had been learned in the auditory modality than in the visual one.

Second, we investigated whether sounds were judged longer than visual stimuli during the testing phase. If the reference memory value of duration is a direct transformation of that value previously held in the accumulator or in working memory

(Gibbon, 1991) then, with crossmodal comparisons, longer visual stimuli should be matched to the auditory internal referent and shorter auditory stimuli should be matched to the visual internal referent. The generalization gradient (i.e., the proportion of identifications of a stimulus as being the standard) should be shifted to the right in the AUD-VIS group and to the left in the VIS-AUD group.

Method

Participants

40 students at the University of Hildesheim participated and were randomly assigned to the four conditions. The mean age of the participants was 24.95 years (*SD* = 4.48). 72.5% of them were women, 27.5% were men. They were paid DEM 10 for participation.

Stimuli and apparatus

The stimuli were either acoustic or visual. Similar to other timing experiments (e.g., Meredith & Wilsoncroft, 1989; Wearden & Bray, 2001; Wearden et al., 1998; Wearden, Pilkington, & Carter, 1999), the acoustic stimuli were pure sinus tones (500 Hz) presented binaurally via headphones, and the visual stimuli were grey rectangles (6 x 12 cm) on a black background presented on a computer monitor screen. The standard duration was 400 ms. Non-standard durations were 100 ms, 200 ms, 300 ms, 500 ms, 600 ms, and 700 ms.

Procedure

During the training phase participants were initially instructed to attend to a 400 ms time interval that was called the *standard duration*. The standard duration was presented either as an auditory or a visual stimulus. After the standard had been presented five times, a temporal generalization task followed (see Figure 1).

This task began with presentation of the standard duration. Five seconds after the standard was presented a qualitative identical stimulus followed with either equal or different duration. The task of the participants was to decide whether the duration of the second stimulus equaled that of the first. Accurate feedback as to performance accuracy was given. The temporal generalization task consisted of 50 trials, separated into five blocks. For each trial there was a nominal probability of $p = .40$ that the presented duration would be the standard, and respectively a probability of $p = .10$ for each of the non-standard comparison values. Duration values were chosen randomly within one block. The inter-trial interval varied from two to ten seconds dependent on individual decision time. The training phase ended with a triple repetition of the standard.

After the training phase, the testing phase began. Participants were repeatedly presented with duration values which were either the same as the standard or different from it. The participants had to decide after *each* presentation whether the actual duration was the standard. Feedback was not provided. The aim was that participants should refer only to duration values represented in long-term memory. Each participant received 200 trials, separated into 20 blocks. Probability of stimulus presentation was the same as in the training phase. The whole experiment lasted on average 38.38 minutes (*SD* = 3.31).

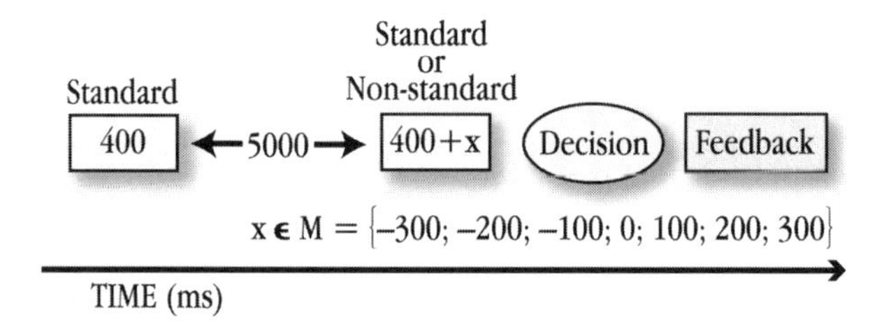

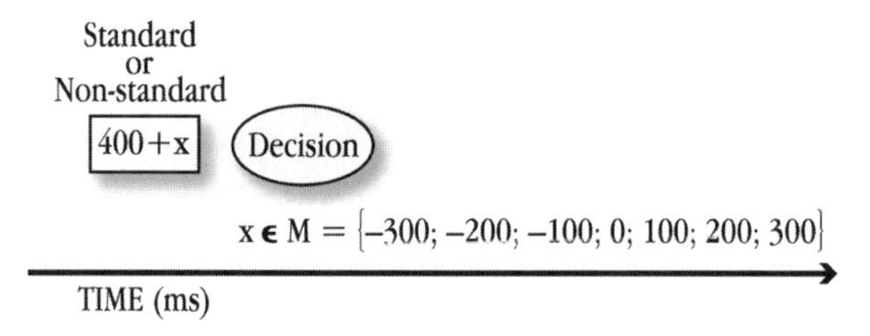

Figure 1: The temporal generalization task of both the training phase and the testing phase of Experiment 1.

Results

Figure 2 shows the distribution of averaged frequencies of *yes* responses (the generalization gradients) of the testing phase of Experiment 1. Response frequencies were divided by frequencies of presentation.

Within both single-modality groups (AUD-AUD, VIS-VIS) the same peak of *yes* responses (400 ms) was shared. Contrary to the results of the single-modality groups and rather unexpectedly, gradients of both crossmodal groups (AUD-VIS, VIS-AUD) show a clear shift to the right of the abscissa. The peak of *yes* responses did not occur at the standard but at the nearest longer comparison duration (500 ms). This means that a stimulus of 500 ms was judged more frequently as the standard than the presented standard itself.

An overall analysis of variance (ANOVA) with "modality of the training phase" and "modality of the testing phase" as between-subjects factors and "stimulus duration" as within-subjects factor produced a significant effect of duration, $F(6, 216)$ =

77.90, $p < .001$, $\eta_p^2 = .684$, and a significant triple-interaction between duration, modality of the training phase, and modality of the testing phase, $F(6, 216) = 11.43$, $p < .001$, $\eta_p^2 = .241$. This result indicates that the apparent difference between the gradients of the single-modality groups and the gradients of the crossmodal groups can be statistically confirmed. Furthermore, the effect of modality of the testing phase was significant, $F(1, 36) = 8.71$, $p = .006$, $\eta_p^2 = .195$.

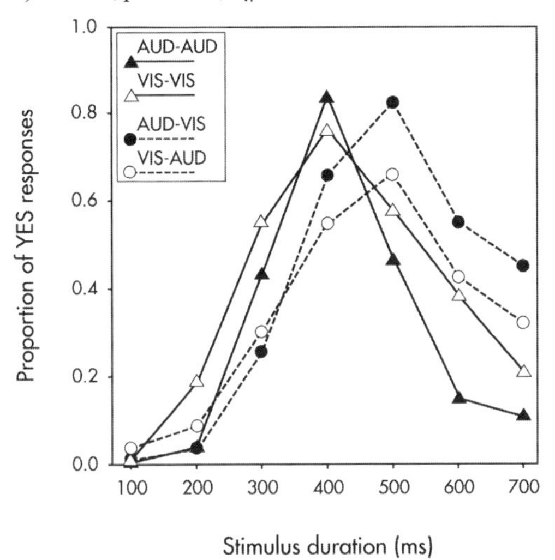

Figure 2: Proportions of *yes* responses in Experiment 1. Triangles and continuous lines indicate data from the single-modality groups, circles and dotted lines indicate data from the crossmodal groups. Adapted from Klapproth (2002, p. 40).[5]

However, the effect of modality of the training phase was not significant, $F(1, 36) = 0.43$, $p = .515$, $\eta_p^2 = .012$. No other F-values were significant.

Accuracy of individuals' judgment was analyzed by entering the proportion of correct responses into an analysis of variance, with "modality of the training phase" (AUD or VIS) and "modality of the testing phase" (AUD or VIS) as between-subjects factors. Mean values and standard deviations for the groups with auditory training phase were 0.73 (*SD* = 0.07), and for the groups with visual training phase 0.68 (*SD* = 0.10), respectively. ANOVA produced no main effect of modality of either experimental phase but a significant interaction between the training phase and the testing phase, $F(1, 36) = 18.02$, $p < .001$, $\eta_p^2 = .334$.

A rightward shift of generalization gradients during repeated presentation of stimuli of changing duration without providing the participants with feedback was reported by Wearden, Pilkington, and Carter (1999) who investigated the type of memory loss of duration values stored in long-term memory. An auditory group and a visual group were tested separately. Generalization gradients shifted progressively to

[5] Reprinted with permission from Elsevier.

the right during the testing phase in the visual group, but only marginally in the auditory group. Wearden et al. (1999) suggested that an arousal-related change in timing processes caused this "subjective lengthening" effect. Consequently, when humans make repeated judgments of the duration of a stimulus without feedback for a period of time, arousal falls and therefore the rate of the pacemaker decreases. In effect, this means that participants need longer stimuli to reach their criterion for the standard duration.

Does the effect of subjective lengthening offer an explanation for the shift effect of the crossmodal conditions? Suppose that subjective lengthening occurred only when the stimuli were presented with the visual mode, but not with the auditory mode. Then, in the AUD-VIS group the gradient should shift to the right because as the testing phase proceeds the rate of the pacemaker decreases. On the other hand, in the VIS-AUD group the rightward shift should be smaller than in the AUD-VIS group.

Whether subjective lengthening occurred within the present study can be examined subsequently by comparison of the proportion of *yes* responses calculated separately for different parts of the testing phase. In order to create a measure of the gradient's asymmetry, the proportions of *yes* responses at duration values longer than the standard were subtracted from the proportions of *yes* responses at duration values shorter than the standard (cf. Wearden et al., 1999). A positive value of this "asymmetry score" indicates a shift of the gradient to the right, a negative value indicates a gradient skewed to the left. For each experiment of this study asymmetry scores were calculated for five blocks including the trials 1–40, 41–80, 81–120, 121–160, and 161–200. Figure 3 shows the progress of rightward asymmetry of the gradients over blocks.

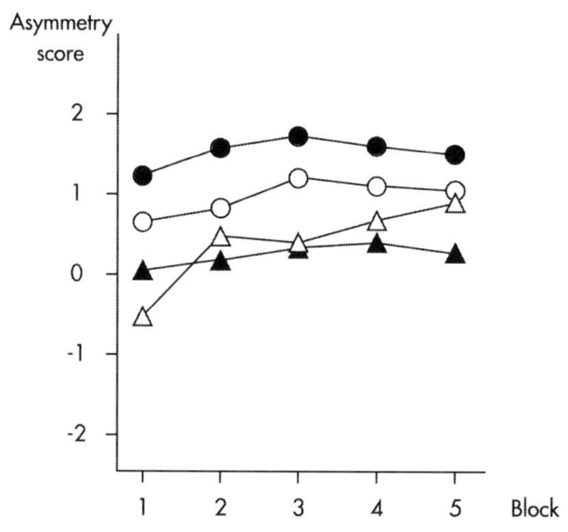

Figure 3: Asymmetry scores plotted against blocks of testing in Experiment 1. Filled triangles: AUD-AUD condition; open triangles: VIS-VIS condition; filled circles: AUD-VIS condition; open circles: VIS-AUD condition.

An analysis of variance with "condition" as the between-subjects factor, "block" as the within-subjects factor, and the asymmetry score as the dependent variable yielded a significant effect of block, $F(4, 144) = 4.46$, $p = .002$, $\eta_p^2 = .11$, and a main effect of condition, $F(3, 36) = 5.65$, $p = .003$, $\eta_p^2 = .32$. The block x condition interaction was not significant, $F(12, 144) = 0.67$, $p = .776$, $\eta_p^2 = .053$.

Discussion

No effect of training modality was found within the testing phase. Therefore, the hypothesis that duration values of auditory stimuli can be better remembered than duration values of visual stimuli has to be rejected. It is probable that providing the participants with feedback during the training phase eliminated the auditory-visual difference of variability in the memory representation of duration. Instead, duration discrimination was dependent on the modality of the testing phase, a result that is comparable to the modality effect found in common discrimination experiments.

Moreover, sounds were not judged longer than lights during the testing phase. The predicted shift of generalization gradients during crossmodal matching only occurred in the AUD-VIS group, but not, as expected, in the VIS-AUD group. In contrast, both crossmodal groups showed a rather uniform shift of the generalization gradient to the right. Thus, participants in the crossmodal groups systematically underestimated presented duration values. This effect occurred *independently* of the direction of the crossmodal comparison. This suggests that the effect does not depend on modality of presentation of stimuli in either experimental phase but rather with the *change* of modality.

As Wearden et al. (1999) reported, temporal generalization gradients shifted progressively to the right with a prolonged testing phase. However, as indicated by the absence of an interaction between block and condition, there were no differences between the conditions regarding the development of the tendency to respond *yes* after stimuli longer than the standard. Furthermore, an underestimation of the presented duration within the crossmodal conditions occurred even during the first 40 trials of the testing phase. Thus, although the phenomenon of "subjective lengthening" occurred, it does not explain the extended shift of the gradients of the crossmodal conditions.

It is obvious that clock-speed differences between auditory and visual modality can not account for the shift of the crossmodal groups in this experiment. Although differences in clock speed should not be neglected it seems that any such differences have been suppressed by a newly discovered effect. In order to confirm this result, I carried out a second experiment.

Experiment 2

To obtain more detailed information about the nature of that "modality-change effect" the temporal generalization task was made more difficult. For this purpose, in the next experiment duration values around the standard were spaced with 50 ms steps instead of 100 ms steps. Thus, the nearest comparison duration value smaller than the standard was 350 ms and the nearest comparison duration value larger than the standard was 450 ms. Consequently, the number of nonstandard comparisons increased from six to eight. With Experiment 2 I focused on the shift effect obtained from Experiment 1. Therefore, only crossmodal conditions were examined with this experiment.

Method

Participants

Twenty students at the University of Hildesheim were randomly allocated to two equal-sized groups. The mean age of the participants was 23.80 years (*SD* = 3.40). 80% were women, 20% were men. DEM 10 was paid for participation.

Stimuli, apparatus, and procedure

Equipment was the same as in Experiment 1. The procedure was identical to that used in Experiment 1, except that the number of non-standard comparisons was raised from six to eight. Thus, the duration values were 100 ms, 200 ms, 300 ms, 350 ms, 400 ms, 450 ms, 500 ms, 600 ms, and 700 ms. In order to maintain the probability of the standard duration ($p = .40$) the standard was presented 16 times during a block, and each of the non-standard durations was presented three times. Therefore, a block was composed of 40 trials. During the training phase participants were presented with only one block, i.e., 40 trials instead of 50 trials in Experiment 1. The testing phase consisted of five blocks (200 trials). Total duration of Experiment 2 was on average 36.95 minutes (*SD* = 7.09).

Results

In Figure 4 the mean proportion of *yes* responses of both crossmodal groups (AUD-VIS, VIS-AUD) is illustrated.

As expected, the gradients of both crossmodal groups show a clear shift to the right. More *yes* responses occurred after presentation of duration values larger than the standard than after presentation of duration values smaller than the standard. Moreover, within both crossmodal groups the peak of *yes* responses did not occur at the standard. In the AUD-VIS group the maximum proportion of *yes* responses shifted from 400 ms to 500 ms, and in the VIS-AUD group the peak of *yes* responses was at 450 ms.

ANOVA with “condition” as between-subjects factor and “stimulus duration” as within-subjects factor produced a significant effect of duration, $F(8, 144) = 29.81$, $p < .001$, $\eta_p^2 = .623$. The main effect of condition, $F(1, 18) = 0.25$, $p = .622$, $\eta_p^2 = .014$, and the interaction, $F(8, 144) = 1.57$, $p = .139$, $\eta_p^2 = .08$, were not significant. These results indicate that the overall proportions of *yes* responses did not differ statistically between both conditions.

As in Experiment 1 the gradient’s shift increased with increasing the duration of the testing phase. The longer the testing phase lasted, the larger was the amount of underestimation of the presented duration. Analysis of variance with “condition” as the between-subjects factor and “block” as the within-subjects factor produced a significant effect of block, F (4, 72) = 6.06, $p < .001$, $\eta_p^2 = .252$, but neither a significant effect of condition, F (1, 18) = 3.27, $p = .087$, $\eta_p^2 = .154$, nor a significant interaction, F (4, 72) = 0.86, $p = .493$, $\eta_p^2 = .046$.

Figure 5 illustrates the progress of asymmetry of the gradients.

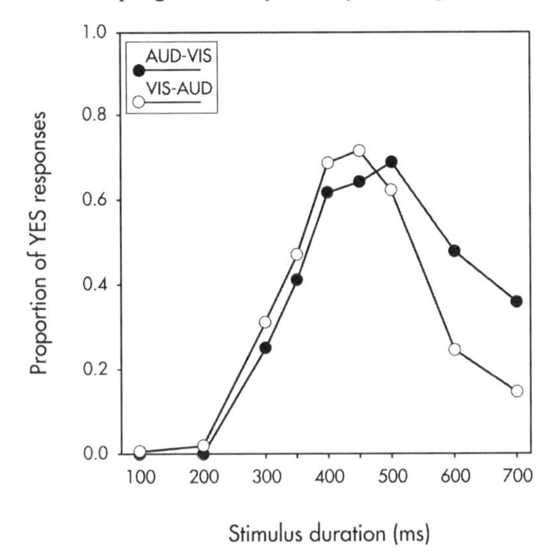

Figure 4: Proportions of *yes* responses in Experiment 2. Filled circles represent data from the AUD-VIS group, open circles represent data from the VIS-AUD group. Adapted from Klapproth (2002, p. 43)[6]

Discussion

Data from the testing phase confirmed the shift effect of Experiment 1. Even when discriminability was weakened because of a change in stimulus spacing a clear shift towards larger duration values occurred within both crossmodal groups. However, statistically significant differences between both crossmodal groups were not obtained. Although the shift of the AUD-VIS group was larger than the shift of the VIS-AUD group, the measure that was used did not produce significance.

As in the previous experiment an increasing tendency to respond with *yes* at longer stimulus durations with a prolonged testing phase was observed. Nevertheless, underestimation took place even within the first 40 trials (block 1) in both conditions because the asymmetry scores were positive (although larger in the AUD-VIS group than in the VIS-AUD group).

[6] Reprinted with permission from Elsevier.

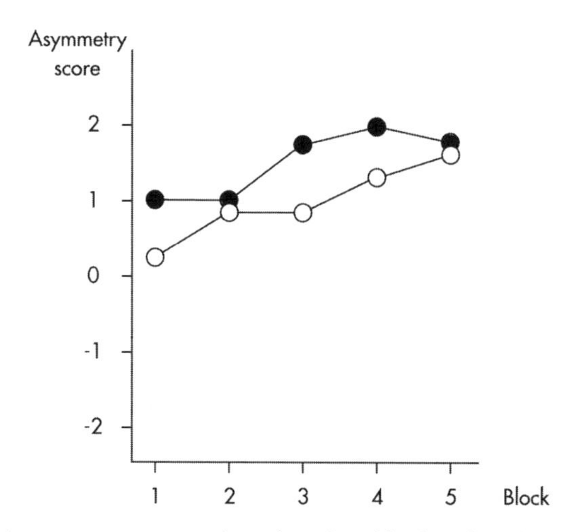

Figure 5: Asymmetry scores plotted against blocks of testing in Experiment 2. Filled circles: AUD-VIS condition; open circles: VIS-AUD condition

Experiment 3

It could be stated that the obtained shift effect is not only due to a change of modality but even possibly due to a change of *any* other stimulus feature. Therefore, the next experiment examined whether variation of *intra-modal* stimulus quality affects the performance in the testing phase. Two conditions where stimulus quality remained constant between training and testing were compared with two conditions in which stimulus quality varied between training and testing. The stimuli were a sinus tone of 500 Hz and white noise.

Method

Participants

20 students were randomly allocated to four equal-sized groups. Mean age of the participants was 25.4 years (*SD* = 4.03). 80% were women, 20 % were men. Each participant received DEM 10 for cooperation.

Stimuli and apparatus

The computer equipment and the duration values were the same as in the previous experiments. The visual stimulus was replaced by a second auditory stimulus, namely white noise.

Procedure

Instructions and the procedure were similar to those used in the previous experiments. During the training phase participants received two blocks (20 trials). The testing phase was composed of 20 blocks resp. 200 trials. The average duration of the experiments was 34.6 minutes (*SD* = 7.89).

Results

Figure 6 shows the gradients of the four conditions.

ANOVA with "condition" (constant versus varying quality) as the between-subjects factor and "duration" as the within-subjects factor yielded a significant effect of duration, $F(6, 108) = 47.15$, $p < .001$, $\eta_p^2 = .724$, and a significant effect of condition, $F(1, 18) = 10.55$, $p = .004$, $\eta_p^2 = .369$. The interaction between duration and condition was not significant, $F(6, 108) = 0.631$, $p = .705$, $\eta_p^2 = .034$.

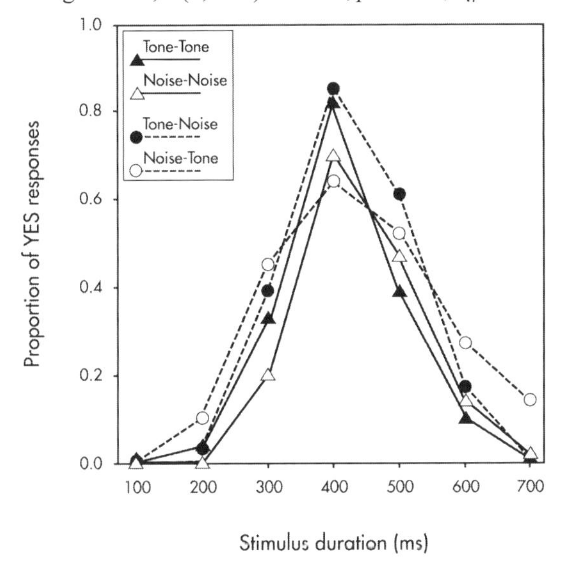

Figure 6: Proportion of *yes* responses in Experiment 3. Triangles and continuous lines indicate data from the single-quality groups, circles and dotted lines indicate data from the cross-quality groups.

Discussion

As expected, changing the quality of the stimulus between the training phase and the testing phase affected the discrimination in the testing phase. Participants in the cross-quality groups responded *yes* more frequently at non-standard comparison durations than participants of the single-quality groups.

However, both groups did not differ regarding the amount of underestimation of the presented duration because there was no shift to the right in the generalization gradients. Apparently, the factors that effected discrimination errors of the cross-quality groups *within one modality* are different from those that produced discrimination errors within *crossmodal* comparison.

It might be the case that in this experiment discrimination errors are due to *changes in the decision criterion*. The gradients of the cross-quality groups were flattened on both sides around the standard; thus the criterion of the cross-quality groups could have been somewhat laxer than in the single-quality groups.

General Discussion

Now I want to suggest a possible cause for the erroneous crossmodal judgments. In order to explain the results of the crossmodal conditions by using the framework of *scalar timing theory* it is necessary to refer to Wearden's (1992) decision rule. This rule states that individuals judge two time intervals as being equal when

$$\frac{abs(s^*-t)}{t} < b^*,$$

where s^* is the memorized standard, t is the perceived duration of the just-presented stimulus, and b^* is the decision criterion.

How, then, could the fact of underestimation of perceived duration be integrated into this formula? Basically, two different possibilities have to be taken into consideration.

First, s^* could have been *increased* with crossmodal comparison. With an increased value of s^* the quotient of the Wearden formula will be minimized and hence the probability of a *yes* response will be maximized if the presented interval is *longer* than the standard. According to this interpretation the value of the memorized standard will grow when the modality changes at the beginning of the testing phase. This implies that the memory representation of the standard is able to change by varying the sensory mode of the timing process. Although changes of the memory representation of time intervals have been discussed frequently (for instance, in connection with the phenomenon of *subjective shortening*; cf. Spetch & Wilkie, 1982; Wearden & Ferrara, 1993), they have been treated commonly as a function of *delay* between learning and retrieval, not as a function of a modality switch.

Second, t could have been *decreased* with crossmodal comparison. Decreasing t will have the same effect as increasing s^*: more *yes* responses will be expected with intervals longer than the standard. This interpretation implies a shortening of perceived duration within the crossmodal conditions which can be attributed to at least two different causes. On the one hand it can be supposed that the speed of emitting pulses from the pacemaker was slowed down. Decelerating the internal clock will reduce the count of accumulated pulses during a given time period, and fewer accumulated pulses correspond to a shorter duration. On the other hand, accumulation of pulses could have been disrupted or delayed, respectively. With a delay of the start of pulse accumulation fewer pulses will have been accumulated during the presented signal. The result will be an underestimation of the signal's duration.

The assumption of modality-dependent changes of the internal clock's speed is not entirely unfounded but it cannot explain the uniform shifts in both crossmodal conditions. Therefore, the interpretation of the shift as an effect of a delayed timing process should be inspected more closely. If a delay of the timing process has occurred within the crossmodal conditions it must have been caused by the change of modality between the training and the testing phase.

But why should the accumulation process be delayed? In a recent article (Klapproth, 2002) I proposed that participants of the crossmodal groups had to switch between the modality of the memorized standard and the modality of the to-be-judged duration. I will now provide some arguments that could elaborate this "modality-switch idea". First, I would like to examine the process of comparison between the reference (or standard) duration (s^*) and the current (or comparison) duration (t) more carefully.

What is the nature of the comparison process?

According to *scalar timing theory* a transfer of interval information from the reference memory to the comparator is needed when the length of a current interval has to be estimated (see Figure 7).

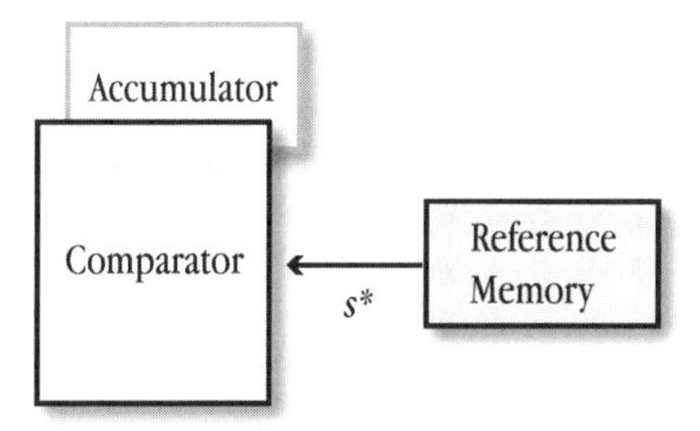

Figure 7: Process of transfer of interval information (s^*) from the reference memory to the comparator (after *scalar timing theory*; e.g., Gibbon & Church, 1984).

Whereas time information stored in memory can be considered as a discrete state, the transfer of time information and its exchange between components of an information processing system is only possible in connection with a material or neural process. If time information has to be transferred from reference memory into working memory for the sake of comparison its discrete state has to be integrated into a neural carrier process (cf. Klix, 1973, p. 63). I assume the easiest way to think about the nature of the transfer process is to consider it as being a reflection of the values of time that have to be transferred.

It is conceivable that the properties of the transfer process are comparable to those of the process that is assumed for the transformation of the duration of a sensory stimulation into a sequence of pulses. The way pulses are getting into memory systems (namely sequentially) should be the same as getting out of them (see figure 8).

The more pulses are transferred, the longer the process should last. Activation of a stored duration value appears to be the same as changing an (inactive) symbol for duration into a real process where processing time represents external time. The static representation of duration in long-term memory turns into a "dynamic mental representation" in which time is an inseparable part of the representation (cf. Freyd, 1987, 1992). Thus, it could be stated that the reference duration (s^*) is transformed from a discrete format (a symbolic value) to an analogous format (the real duration of a process).

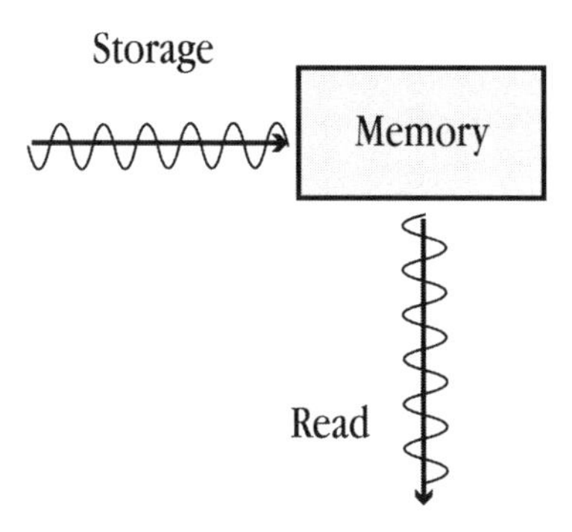

Figure 8: Illustration of the assumed equivalence between the process of accumulation of pulses (stemming from a pacemaker) and the process of reading interval information from reference memory

To express this in more formal terms:

$$S^* = s^* \times t_p,$$

where S* reflects the real time of reading memory, s^* is the number of transferred pulses from memory, and t_p is the time that is required for transferring one pulse,

or

$$S^* = s^* / \lambda_p,$$

where λ_p is the pulse frequency.

Therefore, it can be supposed that comparison between perceived and remembered time is not drawn by computing the numerical difference between two values but, instead, is based on the direct comparison of the duration of two real-time internal processes.

Such an *analogous* comparison process demands that the internal duration from the reference memory and the process of accumulation must occur *in parallel*, that is, they have to start *at the same time*. Then we can imagine the following temporal passage of the comparison process (see Figure 9).

Figure 9 is subdivided into four lines. Line 1 shows a symbol for the physical signal (e.g., a tone or a picture) that is presented in order to be judged by a person.

This signal starts at time P_{ON} (which means onset of the physical signal), and it ceases at P_{OFF} (which means offset of the signal). Assuming some latency of the sensory system the internal duration begins at I_{ON} (with starting the emission of pulses) and ceases at I_{OFF} (line 2). Reading the reference duration starts at R_{ON} and ends at R_{OFF}. To yield optimal comparison results the start of reading the reference duration equals the beginning of the pulse emission. Just as an example, in line 3 the reference duration is a bit shorter than the time of accumulation. Thus, the subject will perceive t longer than S^* ($S^* < t$). If the start of reading the reference duration is delayed (line 4), the signal duration will be underestimated ($S^* > t$) because the process of accumulation (t) is shorter than the reference duration (S^*).

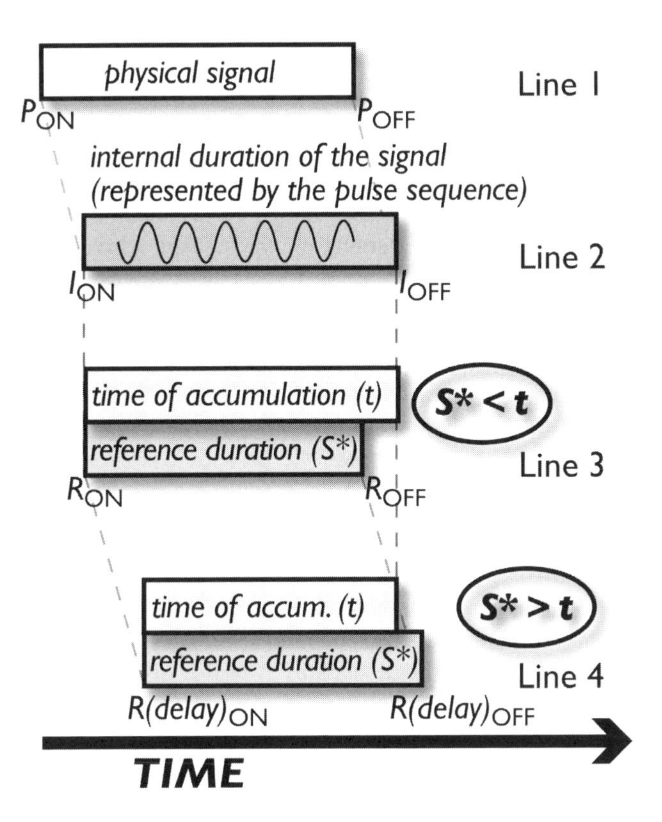

Figure 9: Illustration of the real-time process of the comparison between the duration of an external signal (t) and a memorized reference duration (S^*). See text for details.

This real-time comparison process described above has a resemblance to Kristofferson's "real-time criterion theory" which states among other things that for duration discrimination order information is sufficient (Kristofferson, 1977).

Finally, what of the interval-measure hypothesis? Its rejection obviously does not mean that a measure of a stimulus duration cannot be taken, coded, and stored in memory. The real-time criterion mechanism itself suggests that this is possible and also that the memory representation can be decoded into a real, if internal, time interval. What is rejected is the hypothesis that an order decision about a stimulus duration is based upon a comparison between a coding of the presented stimulus and a representation held in memory (Kristofferson, 1977, p. 116).

What can be concluded from this model? I think it is likely that the empirical result of underestimation of presented duration in the crossmodal conditions could be explained by a delay of accumulation of pulses (t) which is itself caused by a delay of the presence of the internal reference duration (s^*). The switch is able to close and thereby the process of accumulation is able to start only when the reference duration is being activated. The bias obtained in the crossmodal conditions reflects the time the switch has to wait for until it can close.

To recapitulate the question "Why should the accumulation process be delayed?" it could now be answered: Perhaps because the internal standard is delayed. What could be the reason for such a delay?

If, as it happened within Experiment 1 and 2, the standard duration can be better remembered when the modality of learning and the modality of retrieval are the same rather than different, it is reasonable to assume an association between the memorized standard and the modality of its encoding. If this assumption is correct the "encoding specifity principle" (Tulving & Thomson, 1973) should be appliable to the interpretation of the performance in the crossmodal conditions. The *encoding specifity principle* states that recollection of an event, or a certain aspect of it, occurs if and only if properties of the trace of the event are sufficiently similar to the retrieval information (Tulving, 1983) and therefore "the probability of recognition increases with an increase in the number of common features activated at encoding and retrieval" (Jacoby & Craik, 1979, p. 5). Considering the fact that in the crossmodal conditions the number of common stimulus features between training and testing was smaller than in the single-modality conditions it could be supposed that the power of bringing the reference memory into an active state was smaller in the crossmodal conditions than in the single-modality conditions. Possibly, activation of the reference memory is easier or faster if the features of testing and encoding are the same than if they are different. However, at this stage of investigation it remains unclarified whether activation of the reference duration requires an additional process at crossmodal comparison.

Finally, I would like to offer two conclusions that can be drawn from the reported results and their interpretations.

(1) The effect of understanding within crossmodal comparison suggests a connection between sensory modality and temporal memory.

(2) Assuming an analogous comparison process the effect of underestimation seems to be due to a delay of reading the reference memory.

References

Collier, G. L., & Logan, G. (2000). Modality differences in short-term memory for rhythms. *Memory and Cognition, 28*, 529–538.

Freyd, J. J. (1987). Dynamic mental representations. *Psychological Review, 94*, 427–438.

Freyd, J. J. (1992). Dynamic representations guiding adaptive behavior. In F. Macar, V. Pouthas, & W. J. Friedman (Eds.), *Time, action and cognition: Towards bridging the gap* (pp. 309–323). Dordrecht, The Netherlands: Kluwer Academic Publishers.

Gibbon, J. (1991). Origins of scalar timing. *Learning and Motivation, 22,* 3–38.

Gibbon, J., & Church, R. M. (1984). Sources of variance in information processing models of timing. In H. L. Roitblat, T. G. Bever & H. S. Terrace (Eds.), *Animal cognition* (pp. 465–488). Hillsdale, NJ: Erlbaum.

Glenberg, A. M., Mann, S., Altman, L., Forman, T., & Procise, S. (1989). Modality effects in the coding and reproduction of rhythms. *Memory and Cognition, 17*, 373–383.

Goldstone, S., & Goldfarb, J. L. (1964). Auditory and visual time judgment. *The Journal of General Psychology, 70*, 369–387.

Goldstone, S., & Lhamon, W. T. (1974). Studies of auditory-visual differences in human time judgment: 1. Sounds are judged longer than lights. *Perceptual and Motor Skills, 39*, 63–82.

Hocherman, S., & Ben-Dov, G. (1979). Modality-specific effects on discrimination of short empty time intervals. *Perceptual and Motor Skills, 48*, 807–814.

Jacoby, L. L., & Craik, F. I. M. (1979). Effects of elaboration of processing at encoding and retrieval: Trace distinctiveness and recovery of initial context. In L. S. Cermak & F. I. M. Craik (Eds.), *Levels of processing in human memory*. Hillsdale, NJ: Erlbaum.

Klapproth, F. (2002). The effect of study-test modalities on the remembrance of subjective duration from long-term memory. *Behavioural Processes, 59*, 37–46.

Klix, F. (1973). *Information und Verhalten. Kybernetische Aspekte der organismischen Informationsverarbeitung. Einführung in naturwissenschaftliche Grundlagen der Allgemeinen Psychologie.* Berlin: VEB Deutscher Verlag der Wissenschaften.

Kristofferson, A. B. (1977). A real-time criterion theory of duration discrimination. *Perception and Psychophysics, 21*, 105–117.

Meredith, L. S., & Wilsoncroft, W. E. (1989). Time perception: Effects of sensory modality, ambient illumination and intervals. *Perceptual and Motor Skills, 68*, 373–374.

Penney, T. B., Allan, L. G., Meck, W. H., & Gibbon, J. (1998). Memory mixing in duration bisection. In D. A. Rosenbaum, & C. E. Collyer (Eds.), *Timing of behavior: Neural, computational, and psychological perspectives* (pp. 165–193). Cambridge, MA: MIT Press.

Penney, T. B., Gibbon, J., & Meck, W. H. (2000). Differential effects of auditory and visual signals on clock speed and temporal memory. *Journal of Experimental Psychology: Human Perception and Performance, 26*, 1770–1787.

Rousseau, L., & Rousseau, R. (1996). Stop-reaction time and the internal clock. *Perception and Psychophysics, 58*, 434–448.

Sebel, A. J., & Wilsoncroft, W. E. (1983). Auditory and visual differences in time perception. *Perceptual and Motor Skills, 57*, 295–300.

Spetch, M. L., & Wilkie, D. M. (1982). A systematic bias in pigeons' memory for food and light durations. *Behavior Analysis Letters, 2*, 267–274.

Tulving, E. (1983). *Elements of episodic memory*. New York: Oxford University Press.

Tulving, E., & Thomson, D. M. (1973). Encoding specifity and retrieval processes in episodic memory. *Psychological Review, 80*, 352–373.

Wearden, J. H. (1992). Temporal generalization in humans. *Journal of Experimental Psychology: Animal Behavior Processes, 18*, 134–144.
Wearden, J. H. (1999). "Beyond the fields we know...:" Exploring and developing scalar timing theory. *Behavioural Processes, 45*, 3–21.
Wearden, J. H., & Bray, S. (2001). Scalar timing without reference memory? Episodic temporal generalization and bisection in humans. *The Quarterly Journal of Experimental Psychology, 54B*, 289–309.
Wearden, J. H., Edwards, H., Fakhri, M., & Percival, A. (1998). Why "sounds are judged longer than lights": Application of a model of the internal clock in humans. *The Quarterly Journal of Experimental Psychology, 51B*, 97–120.
Wearden, J. H., & Ferrara, A. (1993). Subjective shortening in humans' memory for stimulus duration. *Quarterly Journal of Experimental Psychology, 45B*, 163–186.
Wearden, J. H., Pilkington, R., & Carter, E. (1999). "Subjective lengthening" during repeated testing of a simple temporal discrimination. *Behavioural Processes, 46*, 25–38.

Chapter 6: Sensory and cognitive mechanisms in temporal processing elucidated by a model system approach

Thomas Rammsayer

Abstract

There is preliminary evidence that temporal processing of intervals in the range of seconds or more is cognitively mediated, whereas processing of brief durations below approximately 500 ms appears to be based on processes that are automatic, and, most likely, located at a subcortical level. Although the very nature of the internal clock(s) is still to be discovered, neuropharmacological studies in animals, abnormal findings in temporal processing observed with neurologic patients, and pharmacopsychological experiments in healthy human subjects support the notion that timing mechanisms involved in temporal information processing are modulated by different neurotransmitters in the brain.

The experiments summarized in this paper were motivated by the question about the neurobiological basis of temporal information processing. A second purpose of this series of experiments was to provide additional evidence for the assumption of two different timing mechanisms underlying temporal information processing in the range of milliseconds and seconds, respectively.

Since it does not seem possible to experimentally manipulate a single neurotransmitter system, the most suitable strategy for studying brain-behavior relationships represents the so-called model systems approach. This approach can be described best as a single-behavior-multiple-brain-systems strategy that originates with a well-characterized behavior. By attempting to discover how different neurotransmitter systems in the brain contribute to the specific behavior under investigation, the single-behavior-multiple-brain-systems strategy facilitates inferences from a single behavior to underlying neurobiological processes.

Over the past 15 years, we systematically studied the effects of pharmacologically induced changes in various neurotransmitter systems on temporal processing of durations in the range of seconds and milliseconds. This experimental strategy proved to be highly promising for elucidating mechanisms underlying temporal information processing in humans.

The overall pattern of results suggests that active information processing in working memory is involved in temporal processing of intervals in the range of seconds. Therefore, any treatment that either directly affects or effectively interferes with working memory processes may result in impaired temporal discrimination of longer intervals. Temporal processing of extremely brief intervals in the range of milliseconds, on the other hand, appears to be beyond cognitive control and primarily mediated by dopaminergic activity in the basal ganglia.

Basic concepts of sensory and cognitive timing mechanisms

Performance on time perception and duration discrimination in humans as well as time-related behavior in animals has been often explained by the assumption of a hypothetical internal clock based on neural counting (Allan, Kristofferson, & Wiens, 1971; Church, 1984; Creelman, 1962; Getty, 1975; Gibbon, 1977; Killeen & Weiss, 1987; Meck, 1983; Rammsayer & Ulrich, 2001; Treisman, 1963; Treisman, Faulkner, Naish, & Brogan, 1990). According to this account, a neural pacemaker generates pulses, and the number of pulses relating to a physical time interval is recorded by an accumulator. Thus, the number of pulses counted during a given time interval is the internal representation of this interval. Hence, the higher the clock rate of the neural pacemaker the finer the temporal resolution of the internal clock will be, which is equivalent to more accuracy and better performance on timing tasks. The concept of an internal clock underlying temporal information processing has been a central feature of many theoretical accounts of time perception (Allan, 1992).

As early as 1889, however, Hugo Münsterberg put forward the idea of two distinct timing mechanisms involved in temporal information processing: a sensory mechanism for processing of durations less than one third of a second and a non-sensory mechanism for processing of longer durations. Numerous subsequent theoretical accounts and experimental findings endorsed the notion of two distinct timing mechanisms—one for temporal processing of extremely brief durations in the range of milliseconds and one for processing of longer durations in the range of seconds (e.g., Fraisse, 1984; Michon, 1985; Mitrani, Shekerdjiiski, Gourevitch, & Yanev, 1977; Rammsayer, 1996; Sturt, 1925). For example, Mitrani et al. (1977) showed that LSD and mescaline, substances that strongly affect temporal processing of longer intervals, did not alter duration discrimination of intervals in the range of milliseconds. On the basis of their results, Mitrani et al. (1977) concluded that brief time intervals in the range of milliseconds are processed almost automatically at a lower level of the central nervous system and beyond cognitive control. Similarly, Michon (1975, 1985) argued that temporal information processing of intervals longer than approximately 500 ms is cognitively mediated, while temporal processing of shorter intervals are predicted to be highly perceptual in nature, parallel, and not accessible to cognitive control.

There is converging experimental evidence that temporal processing in the range of seconds is cognitively mediated. Cognitive influences have been amply documented in several studies showing that subjects' judgments of durations are a function of the amount of mental content (Frankenhaeuser, 1959), complexity of information (Ornstein, 1969), processing effort (Burnside, 1971), and experience of change (Block & Reed, 1978). Although these studies primarily applied a retrospective timing paradigm, i.e., subjects did not know in advance that they will be asked to make a judgment on time, the involvement of cognitive processes has also been established for prospective timing paradigms in which subjects knew before performing a given task that a time-related activity will be required (for a comprehensive review of retrospective and prospective duration judgments see Block & Zakay, 1997). For example, in a series of experiments with human subjects, Rammsayer and Lima (1991), using a dual-task paradigm, found that temporal processing of brief auditory intervals ranging from 50 to 98 ms was not affected by experimentally induced high memory load. In the same study, however, high memory load caused a marked impairment of duration discrimination performance with longer intervals ranging from 1 to 2 s. Furthermore, Abel (1972a, 1972b) investigated subjects' ability to discriminate a difference in the duration of auditory intervals with base durations ranging from 0.16 to 960 ms. Her results suggested that the theory best describing the data over the largest range of base durations was the neural counting model proposed by Creelman (1962). Most interestingly, however, the neural counting model appeared to fail for base durations longer than approximately 100 ms. Additional experimental evidence for the involvement of memory processes in temporal processing in the range of seconds was provided by Fortin and her colleagues (Fortin & Breton, 1995; Fortin & Rousseau, 1987; Fortin, Rousseau, Bourque, & Kirouac, 1993) who showed that nontemporal processing in working memory interferes selectively with estimation of time. Finally, in one of the most recent accounts of animal timing (Staddon & Higa, 1999), decay of memory strength serves the role of an internal timing mechanism. With this approach, specific temporal intervals are associated with specific amounts of decay in memory.

Attention represents another cognitive process that has often been reported to modulate the efficiency of timing performance (Block, 1990; Block, this volume, chapter 3; Brown, 1985, 1997; Zakay, 1998). Most direct evidence for the critical influence of attention on temporal information processing was provided by Grondin and Macar (1992) who adapted a task, usually employed for attention-operating characteristic analysis, to temporal discrimination. Before each trial, the subject was asked to distribute the amount of attention to each of two tasks to be performed simultaneously. With this approach, it could be shown that more attention allocated to time resulted in better performance on temporal discrimination (Grondin & Macar, 1992; Macar, Grondin, & Casini, 1994).

The attentional-gate model proposed by Block and Zakay (Block & Zakay, 1996, 1997; Zakay, 2000; Zakay & Block, 1996, 1997; Zakay, Block, & Tsal, 1999) provides a conceptual framework to account for attentional effects on temporal information processing. This most recent model integrates concepts derived from traditional internal-clock models (e.g., Creelman, 1962; Treisman, 1963) and cognitive models of timing (e.g., Brown, 1997; Gibbon & Church, 1984). Basically, the attentional-gate model assumes the existence of a pacemaker which emits pulses that are accumulated by a cognitive counter. On their way to the counter, all pulses must pass through a cognitive gate which is controlled by the amount of attentional resources allocated to temporal information processing. The gate opens more widely or more frequently as more attention is paid to time which, in turn, will result in higher temporal accuracy.

Since any sensory stimulus consists of nontemporal information presented for a given duration, cognitive timing of differences between two intervals requires a subject to divide attention between temporal and nontemporal information processing (Thomas & Weaver, 1975; Zakay & Block, 1996, 1997). For example, when performing a duration discrimination task, processing of nontemporal information interferes with processing of temporal information and, thus, may decrease task accuracy. From this perspective, the more attentional ressources can be allocated to the processing of temporal information, the better temporal accuracy will be.

The model systems approach

Although the experiments summarized in this paper were primarily motivated by the question about the neurobiological basis of temporal information processing, a second purpose of this series of experiments was to provide additional pharmacological evidence for the assumption of two different timing mechanisms underlying temporal information processing in the range of milliseconds and seconds, respectively. To achieve this goal, a so-called model systems approach was applied (cf., Solomon, 1986).

When studying brain-behavior relationships, the researcher will be faced with the problem that manipulation of a single neurotransmitter system may also cause significant changes in levels of activity of other neurotransmitter systems. Since it does not seem possible to experimentally manipulate a single neurotransmitter system without affecting other ones, so-called single-transmitter-multiple-behavior strategies do not appear to be appropriate for elucidation of brain-behavior relationships. Unlike single-transmitter-multiple-behavior strategies, a model systems approach (Kandel, 1976; Thompson, 1976) represents a more suitable strategy for studying brain-behavior relationships. The model systems approach can be described best as a single-behavior-multiple-brain-systems strategy that originates with a well-characterized behavior. By attempting to discover how different neurotransmitter systems in the brain contribute to the specific behavior under investigation, the single-behavior-multiple-brain-systems strategy facilitates inferences from a single behavior to underlying neurobiological processes (Solomon, 1986).

Furthermore, increasing evidence for the neurochemical coding of behavior points to the significance of pharmacopsychological or neuropharmacological approaches within the field of cognitive neurosciences (Russel, 1987). In line with the single-behavior-multiple-brain-systems strategy, the ultimate aim of these approaches is to discover neurochemical brain systems that are mediating specific human behavior. This can be achieved by utilizing the specific pharmacological actions and action mechanisms of drugs for better understanding of the neurobiological basis of behavioral processes (Janke, 1983). In this respect, the present paper represents a cognitive-neuroscience approach using drugs as research tools for elucidation of mechanisms underlying information processing.

Applying a single-behavior-multiple-brain-systems strategy to human temporal information processing, we could show that pharmacological interventions by means of dopamine (DA)-receptor antagonists, the benzodiazepine midazolam, and the noradrenaline-reuptake inhibitor reboxetine effectively modulate timing performance in humans, whereas the serotoninergic (Rammsayer, 1989a) and cholinergic (Rammsayer, 1999) neurotransmitter systems as well as glutamatergic NMDA receptors (Rammsayer, Groh, & Rodewald, 1998) did not appear to play a crucial role in human temporal information processing. How these neuropharmacological findings helped to elucidate distinct timing mechanisms underlying temporal information processing in the range of milliseconds and seconds, respectively, will be depicted in the present contribution.

Psychophysical assessment of timing performance

Over the past 10 years, we systematically investigated the effects of pharmacologically induced changes in various neurotransmitter systems on temporal information processing of durations in the range of seconds and milliseconds. To assess duration discrimination performance in the range of seconds and milliseconds, in all our neuropharmacological studies, difference thresholds were determined in relation to a 1,000-ms and a 50-ms base duration (i.e., standard interval), respectively. The choice of both these base durations was motivated by the notion that the hypothetical shift from one timing mechanism to the other may be found at an interval duration somewhere between 100 and 500 ms (cf., Abel, 1972a, 1972b; Buonomano & Merzenich, 1995; Michon, 1985; Münsterberg, 1889; Rammsayer, 1996). Furthermore, when participants are asked to compare time intervals, many of them count out the required number of seconds. Since explicit counting becomes a useful timing strategy for intervals longer than approximately 1,200 ms (Grondin, Meilleur-Wells, & Lachance, 1999), the "long" base duration was chosen not to exceed this critical value.

An experimental session consisted of one block of duration discrimination in the range of milliseconds and one block of duration discrimination in the range of seconds. The order of blocks was counterbalanced across participants. Each trial consisted of two auditory stimuli, a constant standard interval and a variable comparison interval. The auditory stimuli, separated by an inter-stimulus interval of 900 ms, were presented through headphones. The participant's task was to decide which of the two intervals was longer and to indicate his decision by pressing one of two designated keys on the keyboard. After each response, visual feedback was displayed on the monitor screen.

For both temporal discrimination tasks, order of presentation for the standard interval and the comparison interval was randomized and balanced, with each interval being presented first in 50% of the trials. For quantification of performance on duration discrimination, an adaptive psychophysical procedure was used. "Adaptive" means that stimulus presentation on any given trial is determined by the preceding set of stimuli and responses. Therefore, the comparison interval is varied in duration from trial to trial depending on the participant's previous response. Correct responding resulted in a decrease of the duration of the comparison interval and incorrect responses made the task easier by increasing the duration of the comparison interval. Depending on the adaptive rule chosen to change the duration of the comparison interval, the psychophysical procedure converges to a specific duration of the comparison interval required for a specific level of performance. In the experiments reported here, either the transformed up-down method using a 2-step rule (Levitt, 1971) or the weighted up-down method (Kaernbach, 1991) were applied to determine the 70.7%-

or 75%-difference threshold, respectively, as an indicator of performance on duration discrimination (see Rammsayer, 1992). The difference threshold represents the difference in duration between the standard and the comparison interval required to produce the specified percentage of correct responses. Therefore, better temporal discrimination is indicated by smaller threshold values.

Effects of dopamine receptor blockade on temporal information processing

Neuropharmacological studies in animals suggest that the rate of the hypothesized internal clock used for temporal information processing is positively related to the effective level of brain DA. Based on their results on timing behavior in rats, Church and his colleagues (e.g., Church, 1984; Maricq & Church, 1983; Meck, 1983, 1996) provided first evidence for the notion that DA agonists such as methamphetamine increase clock speed whereas DA antagonists such as haloperidol decrease the mean rate of the pacemaker of the internal clock.

There also is some evidence from human studies pointing to the detrimental effect of haloperidol on temporal information processing. Clinical studies on haloperidol have found significant deficits in tests of duration discrimination with base durations in the range of seconds both in schizophrenic patients and patients with Tourette's syndrome without neurologic disease, whereas untreated patients showed no impairment of temporal processing (Goldstone & Lhamon, 1976; Goldstone, Nurnberg, & Lhamon, 1979). A pharmacological study with healthy male volunteers (Rammsayer, 1989b) showed that duration discrimination in the range of milliseconds was significantly impaired by haloperidol. These findings may point to the significance of dopaminergic activity in human temporal information processing.

Therefore, a first series of experiments was designed to investigate the effects of dopaminergic modulation on temporal information processing in healthy human subjects by directly contrasting performance on duration discrimination of intervals in the range of seconds and milliseconds (Rammsayer, 1993, 1997a, 1997b). The major focus of these experiments was on the critical role of D2 receptor activity for temporal information processing. Because application of D2 receptor agonists in humans is followed by considerable nausea and other adverse reactions, only D2 receptor antagonists were chosen as a research tool for inducing specific changes in D2 receptor activity.

With regard to the pharmacological dissociation of sensory and cognitive mechanisms involved in temporal processing the comparison of two D2 receptor blockers, haloperidol and remoxipride, proved to be most informative. These two D2 receptor blockers were chosen due to their different pharmacological properties on the two major DA brain systems, the mesostriatal and the mesolimbocortical DA system. The mesostriatal DA system consists of dopaminergic neurons projecting from the substantia nigra to the striatal complex, while the neurons of the mesolimbocortical DA system originate in the ventral tegmental area and the medial substantia nigra. Unlike neurons of the mesostriatal DA system, neurons of the mesolimbocortical

DA system primarily project to limbic, allocortical, and neocortical areas (Björklund & Lindvall, 1986). There are also functional differences between both these major DA systems. Mesolimbocortical DA neurons play a crucial role in working-memory processes, associative learning, locomotor activity, active avoidance, and incentive-reward motivation. The mesostriatal DA system, on the other hand, is primarily involved in motor response activation, execution of learned motor programs, and sequencing of behavior. While haloperidol blocks D2 receptors in all DA brain systems, remoxipride primarily blocks D2 receptors of mesolimbocortical DA neurons (Gerlach & Casey, 1990).

In two double-blind studies, either 3 mg of haloperidol, 150 mg of remoxipride, or placebo were administered in a single oral dose (Rammsayer, 1993, 1997a). The selected doses of haloperidol and remoxipride were equated with regard to their antipsychotic potencies (see Rammsayer, 1997a). Performance on temporal processing of intervals in the range of seconds was significantly impaired by both drugs as compared with placebo. On the other hand, only haloperidol produced a significant decrease in performance on temporal processing of intervals in the range of milliseconds as compared with placebo and remoxipride, whereas remoxipride and placebo groups did not differ significantly.

Within the neuropharmacological framework, the differential effects of haloperidol and remoxipride on temporal processing of very brief intervals may reflect the different pattern of the pharmacological action of both drugs. Haloperidol, like most traditional neuroleptics, typically causes unwanted motor side-effects, while the pharmacological profile of remoxipride is that of a so-called "atypical" antipsychotic drug with a low potential for extrapyramidal side-effects (Ögren et al., 1984; Seeman, 1990). An explanation for this relatively low potential for motor side-effects seen with remoxipride may be its more pronounced action on D2 receptors located in the mesolimbocortical areas of the brain than on those in the striatum (Köhler, Hall, Magnusson, Lewander, & Gustafsson, 1990). While the antipsychotic effect of neuroleptics, whether typical or atypical with respect to motor side-effects, appears to depend on their common ability to block D2 receptors in mesolimbocortical areas (e.g., Crow, Deakin, & Longdon, 1977; Seeman, 1987), the differential effects on temporal processing of intervals in the range of milliseconds observed with haloperidol and remoxipride suggest that the pharmacological property to modulate the internal timing mechanism is likely to depend on DA antagonistic effects on the basal ganglia, i.e., the neuroanatomical region where the extrapyramidal motor side-effects, such as parkinsonism, dystonia, akathisia, and tardive dyskinesia, are mediated. Since, if mesolimbocortical DA activity were involved in temporal processing of very brief intervals, performance should have been affected by the typical (haloperidol) as well as the atypical (remoxipride) drug. As this was not found in both studies, the pronounced deteriorating effect of haloperidol on timing of brief intervals as compared to placebo and remoxipride indicates that the timing mechanism underlying temporal processing of intervals in the range of milliseconds depends on D2 receptor activity in the basal ganglia.

The tendency for haloperidol to produce extrapyramidal motor effects more efficiently than remoxipride only ambigously supports the argument that intervals in the millisecond range are processed in the basal ganglia as opposed to other dopaminergic brain regions such as the prefrontal cortex or the nucleus accumbens. Additional converging evidence for the involvement of the basal ganglia in the timing of extremely brief intervals is provided by clinical studies on Parkinson's disease. Patients suffering from Parkinson's disease are characterized by extremely low levels of DA activity in the basal ganglia (Hornykiewicz, 1972). In a study comparing Parkinson patients with age-matched healthy controls, Artieda, Pastor, Lacruz, and Obeso (1992) found that temporal discrimination in the range of milliseconds was significantly impaired in the former group. Similar results are reported by Rammsayer and Classen (1997). Furthermore, accuracy and precision of timing of self-paced movement in Parkinson patients has been shown to vary as a function of DA activity in the basal ganglia (O'Boyle, Freeman, & Cody, 1996). These findings also support the hypothesis that the integrity of the basal ganglia is crucial for faultless and unimpaired processing of temporal information in the range of milliseconds.

A basic premise of cognitive models of temporal processing is that the experienced duration of an interval is a function of the information put into short-term memory (McGrath & Kelly, 1986). Convincing experimental evidence for the involvement of working memory in temporal processing of longer intervals has been provided by Fortin and her colleagues (Fortin & Breton, 1995; Fortin et al., 1993). In a series of highly sophisticated experiments, these authors showed that processing in short-term memory interferes selectively with estimation of intervals in the range of seconds. Furthermore, numerous behavioral and pharmacopsychological studies show that working-memory processes are modulated by dopaminergic mesolimbocortical projections to the prefrontal cortex (e.g., Brozoski, Brown, Rosvold, Goldman, 1979; Goldman-Rakic, 1996; Kimberg, D'Esposito, & Farah, 1997; Luciana, Depue, Arbisi, & Leon, 1992; Müller, von Cramon, & Pollmann, 1998; Park & Holzman, 1993; Sawaguchi, Matsumara, & Kubota, 1990). Accordingly, the deteriorating effects of haloperidol and remoxipride on temporal processing of intervals in the range of seconds may have been caused by drug-induced impairment of memory processes. Such an interpretation is supported by the finding that haloperidol as well as remoxipride adversely affected memory processes in healthy volunteers. For example, 100 mg of remoxipride caused a decrease in performance in memory tasks (Mattila, Mattila, Konno, Saarialho-Kere, 1988) and, similarly, 3 mg of haloperidol caused pronounced deficits in cognitive functioning in healthy human subjects (King, 1993; McClelland, Cooper, & Pilgrim, 1990; Squitieri, Cervone, & Agnoli, 1977).

Thus, from a pharmacopsychological point of view, there is converging evidence for the assumption that the impairing effect of haloperidol and remoxipride on temporal processing of intervals in the range of seconds is very likely brought about by blockade of D2 receptors in the mesolimbocortical DA system. This is because both the typical neuroleptic haloperidol as well as the atypical neuroleptic remoxipride exert DA antagonistic effects in this region of the brain.

Taken together, although the involvement of the basal ganglia in motor activity is by far the most striking of their functions, the outcome of both studies (Rammsayer, 1993, 1997a) strongly suggests that the basal ganglia also play an important role in temporal processing of brief durations in the range of milliseconds. Furthermore, temporal processing of intervals in the range of seconds is dependent on cognitive functioning and, therefore, pharmacological treatment with D2 receptor blockers may produce pronounced deficits in timing performance due to impairment of memory functions mediated by the mesolimbocortical DA system. Converging evidence for the assumption that the timing mechanism underlying temporal processing of longer intervals is functionally different from the one underlying temporal processing of extremely brief intervals comes from a study on the effects of body core temperature and brain DA activity on timing processes in humans (Rammsayer, 1997b).

Effects of benzodiazepine-induced memory impairment on temporal information processing

To provide some additional evidence for the notion of two distinct timing mechanisms underlying temporal processing of intervals in the range of seconds and milliseconds, in a series of two experiments (Rammsayer, 1994, 1999), the effect of benzodiazepine-induced memory impairment on duration discrimination performance was studied. As a research tool to investigate the effects of pharmacologically induced non-dopaminergic memory impairment on temporal information processing, the benzodiazepine midazolam was used. Midazolam is a short-acting, water-soluble benzodiazepine that possesses marked memory impairing effects (File, 1992; File, Skelly, & Girdler, 1992; Rammsayer, 1999; Rammsayer, Rodewald, & Groh, 2000; Subhan & Hindmarch, 1983).

Since the outcome of the experiments with the D2 receptor antagonists haloperidol and remoxipride suggested that temporal processing of longer intervals appears to be cognitively mediated and to be dependent on working-memory processes, we would predict that performance on duration discrimination with a 1-s base duration would also be adversely affected by non-dopaminergic memory-impairing pharmacological treatment. Unlike performance on duration discrimination with a 1-s base duration, performance on duration discrimination with a 50-ms base duration should not be expected to deteriorate because temporal processing of intervals in the range of milliseconds does not involve memory processes.

In two double-blind placebo-controlled experiments either 15 mg (Rammsayer, 1994) or 11 mg (Rammsayer, 1999) of midazolam were administered in a single oral dose. In both studies, temporal processing of intervals in the range of seconds was markedly impaired under midazolam as compared with placebo, while temporal processing of intervals in the range of milliseconds was affected by neither 15 nor 11 mg of midazolam.

Since benzodiazepines are known to directly affect memory functions, the observed decrease in performance on duration discrimination with a base duration of 1 s under midazolam may be due to its impairing effect on memory processes associated with temporal information processing. In this respect, reduced performance integrity can be considered a consequence of pharmacologically induced deterioration of memory functions that result in a more variable cognitive representation of the durations to be compared and, thus, significantly decreasing task accuracy. On the other hand, the lack of a benzodiazepine-induced effect on performance on duration discrimination with a base duration of 50 ms is consistent with the assumption that temporal processing of extremely brief intervals in the range of milliseconds is beyond cognitive control and independent of memory processes (Michon, 1985; Rammsayer & Lima, 1991).

This pattern of results supports the general notion of two distinct timing mechanisms underlying temporal processing of intervals in the range of milliseconds and seconds, respectively, as derived from the outcome of the studies on the effects of D2-receptor antagonists on timing performance. While the timing mechanism for processing of extremely brief intervals seems to be highly sensory in nature and located at a subcortical level of the central nervous system, the one associated with temporal processing of longer intervals primarily involves memory processes.

Differential effects of noradrenergic modulation on temporal processing of intervals in the range of seconds and milliseconds

Since noradrenaline is implicated in the neuronal modulation of attention (Berridge, Arnsten, & Foote, 1993; Coull, 1998; Coull, Frith, Dolan, Frackowiak, & Grasby, 1997), studying the effects of noradrenergic activity in the brain on temporal processing appears to be a worthwhile effort in order to explore the significance of attentional processes for temporal discriminations in humans. Attention has often been reported to modulate the efficiency of timing performance (e.g., Block, 1990; Block, this volume, chapter 3; Brown, 1985, 1997; Grondin & Macar, 1992; Zakay, 1998) and it could be shown that more attention allocated to time resulted in better performance on temporal discrimination (Grondin & Macar, 1992; Macar et al., 1994). Therefore, an experiment (Rammsayer, Hennig, Haag, & Lange, 2001) was designed to investigate the effects of the specific noradrenaline reuptake inhibitor reboxetine on temporal information processing of intervals in the range of seconds and milliseconds.

Performance on temporal processing of longer intervals was significantly improved by a single oral dose of 2 mg of reboxetine as compared to placebo, whereas the improvement observed with a 4-mg dose just failed to reach statistical significance. There was, however, no effect of enhanced noradrenergic activity on temporal processing of very brief intervals. This differential effect of noradrenergic activity on temporal processing of intervals in the range of seconds and milliseconds cannot be explained by pacemaker-related timing variance but rather reflects attentional modulation of timing performance (Rammsayer et al., 2001). Such an interpretation is consistent with the attentional-gate model. Cognitive timing of differences between two intervals requires a subject to divide attention between temporal and nontemporal information processing (Thomas & Weaver, 1975; Zakay & Block, 1996, 1997). When performing a duration discrimination task, processing of nontemporal information interferes with processing of temporal information and, thus, may decrease task accuracy. Therefore, nontemporal information represents a distracting stimulus during temporal processing. Under this condition, noradrenergic activation enabled our subjects to focus their attention on task-relevant (temporal) information and, thus, attenuated the influence of distracting (nontemporal) information.

Results from animal studies provided evidence that noradrenergic activation helps to focus attention on task-relevant information by attenuating the influence of distracting stimuli (Robbins, 1984) whereas noradrenaline depletion, in rodents causes increased distractibility (Carli, Robbins, Evenden, & Everitt, 1983; Roberts, Price, & Fibiger, 1976). Furthermore, in non-human primates, noradrenergic alpha2-receptor agonists, such as clonidine or guanfacine, improved working-memory performance (Arnsten & Goldman-Rakic, 1985; Cai, Ma, Xu, & Hu, 1993; Schneider & Kovelowski, 1990). Similarly, in human patients suffering from dementia of the frontal type or attention-deficit hyperactivity disorders, alpha2-receptor agonists appear to remediate specific attentional functions via frontal cortex (Arnsten, Steere, & Hunt, 1996; Coull, Hodges, & Sahakian, 1996). In addition, pharmacopsychological studies suggested that any pharmacological treatment that interferes with maintaining the activation of memory units or attentional processes in working memory may interfere with temporal processing of longer intervals (Rammsayer, 1999). Based on these findings, the performance-enhancing effect of the noradrenaline reuptake inhibitor reboxetine on temporal processing of longer intervals may be due to its beneficial effects on prefrontal cortex functions such as maintaining working-memory performance under distracting conditions.

Unlike temporal processing of time intervals in the range of seconds, temporal processing of extremely brief intervals below approximately 500 ms appears to be beyond cognitive control and based on processes located at a subcortical level (Michon, 1985; Mitrani et al., 1977; Rammsayer, 1999; Rammsayer & Lima, 1991). Most recently, Mattes and Ulrich (1998) showed that performance on duration discrimination of intervals with base durations up to 300 ms was not influenced by changes in directed attention. Furthermore, their data suggested that attention does not affect the internal-clock mechanism per se. These findings may account for the differential effects of reboxetine on temporal processing of intervals in the range of seconds and milliseconds: while processing of longer durations can, at least partly, be considered a function of directed attention as suggested by the attentional-gate model, temporal discrimination of extremely brief durations seems to be less sensitive to pharmacologically induced changes in directed attention.

These findings provide additional converging evidence for the notion that temporal processing of intervals in the range of seconds is based on working-memory processes including aspects of directed attention. Processing of intervals in the range of milliseconds, on the other hand, is mediated by subcortical processes beyond cognitive control and not responsive to changes in noradrenergic activity.

Conclusions

The outcome of the pharmacopsychological studies presented above suggests that temporal processing of intervals in the range of seconds is based on working-memory processes including aspects of directed attention. Accordingly, any pharmacological treatment that interferes with maintaining the activation of memory units or attentional processes in working memory appears to interfere with temporal processing of longer intervals. This interpretation is corroborated by results from dual-task studies (e.g., Fortin & Breton, 1995; Fortin et al., 1993; Rammsayer & Lima, 1991). All these pharmacopsychological and experimental findings are consistent with the general notion, that any treatment that either directly affects or effectively interferes with active information processing in working memory, results in impaired temporal processing of longer intervals.

Unlike temporal processing of longer intervals, temporal processing of extremely brief intervals in the range of milliseconds appears to be beyond cognitive control, highly sensory in nature, and primarily modulated by DA activity in the basal ganglia.

A most recent reanalysis of the neuropharmacological experiments described above (Hellström & Rammsayer, 2002), suggests that the timing mechanisms underlying temporal processing of intervals in the range of milliseconds and seconds are not completely independent of each other. The sensory ("biological") timing mechanism, which is associated with the timing of very brief intervals, also assists the cognitive timing mechanism for processing of longer intervals. Thus, Hellström and Rammsayer (2002) propose a preliminary model with two internal clocks, one for brief and long intervals, and one for longer intervals only. For the longer intervals, the readings of the two clocks are weighted together in such manner as to improve duration discrimination or timing by minimizing error variance. Additional neuropharmacological studies are needed to further elaborate this model and to assess in more detail the effects of various pharmacological substances on timing performance.

References

Abel, S. M. (1972a). Discrimination of temporal gaps. *Journal of the Acoustical Society of America, 52*, 519–524.

Abel, S. M. (1972b). Duration discrimination of noise and tone bursts. *Journal of the Acoustical Society of America, 51*, 1219–1223.

Allan, L. G. (1992). The internal clock revisited. In F. Macar, V. Pouthas & W. J. Friedman (Eds.), *Time, action and cognition. Towards bridging the gap* (pp. 191–202). Dordrecht, The Netherlands: Kluwer Academic Publishers.

Allan, L. G., Kristofferson, A. B., & Wiens, E. W. (1971). Duration discrimination of brief light flashes. *Perception and Psychophysics, 9*, 327–334.

Arnsten, A. F. T., & Goldman-Rakic, P. S. (1985). Alpha-2 adrenergic mechanisms in prefrontal cortex associated with cognitive decline in aged nonhuman primates. *Science, 230*, 1273–1276.

Arnsten, A. F. T., Steere, J. C., & Hunt, R. D. (1996). The contribution of alpha2-noradrenergic mechanisms to prefrontal cortical cognitive function. *Archives of General Psychiatry, 53*, 448–455.

Artieda, J., Pastor, M. A., Lacruz, F., & Obeso, J. A. (1992). Temporal discrimination is abnormal in Parkinson's disease. *Brain, 115*, 199–210.

Berridge, C. W., Arnsten, A. F. T., & Foote, S. L. (1993). Noradrenergic modulation of cognitive function: clinical implications of anatomical, electrophysiological and behavioural studies in animal models. *Psychological Medicine, 23*, 557–564.

Björklund, A., & Lindvall, O. (1986). Catecholaminergic brain stem regulatory systems. In American Physiological Society (Ed.), *Handbook of Physiology. Section 1. The Nervous System. Vol. IV. Intrinsic regulatory systems of the brain* (pp. 155–235). Bethesda, MD: American Physiological Society.

Block, R. A. (1990). Models of psychological time. In R. A. Block (Ed.), *Cognitive models of psychological time* (pp. 1–35). Hillsdale, NJ: Erlbaum.

Block, R. A., & Reed, M. A. (1978). Remembered duration: evidence for a contextual change hypothesis. *Journal of Experimental Psychology: Learning and Memory, 4*, 656–665.

Block, R. A., & Zakay, D. (1996). Models of psychological time revisited. In H. Helfrich (Ed.), *Time and mind* (pp. 171–195). Seattle, WA: Hogrefe & Huber Publishers.

Block, R. A., & Zakay, D. (1997). Prospective and retrospective duration judgments: a meta-analytic review. *Psychonomic Bulletin and Review, 4*, 184–197.

Brown, S. W. (1985). Time perception and attention: The effects of prospective versus retrospective paradigms and task demands on perceived duration. *Perception and Psychophysics, 38*, 115–124.

Brown, S. W. (1997). Attentional resources in timing: interference effects in concurrent temporal and nontemporal working memory tasks. *Perception and Psychophysics, 59*, 1118–1140.

Brozoski, T. J., Brown, R. M., Rosvold, H. E., & Goldman, P. (1979). Cognitive deficit caused by regional depletion of dopamine in prefrontal cortex of rhesus monkey. *Science, 205*, 929–931.

Buonomano, D. V., & Merzenich, M. M. (1995). Temporal information transformed into a spatial code by neural network with realistic properties. *Science, 267*, 1028–1030.

Burnside, W. (1971). Judgment of short time intervals while performing mathematical tasks. *Perception and Psychophysics, 9*, 404–406.

Cai, J. X., Ma, Y., Xu, L., & Hu, X. (1993). Reserpine impairs spatial working memory performance in monkeys: reversal by the alpha2-adrenergic agonist clonidine. *Brain Research, 614*, 191–196.

Carli, M., Robbins, T. W., Evenden, J. L., & Everitt, B. J. (1983). Effect of lesions to ascending noradrenergic neurons on performance of a 5-choice serial reaction task in rats: Implications for theories of dorsal noradrenergic bundle function based on selective attention and arousal. *Behaviour and Brain Research, 9*, 361–380.

Church, R. M. (1984). Properties of the internal clock. In J. Gibbon & L. G. Allan (Eds.), *Timing and Time Perception* (pp. 566–582). New York: New York Academy of Sciences.

Coull, J. T. (1998). Neural correlates of attention and arousal: insights from electrophysiology, functional neuroimaging and psychopharmacology. *Progress in Neurobiology, 55*, 343–361.

Coull, J. T., Frith, C. D., Dolan, R. J., Frackowiak, R. S. J., & Grasby, P. M. (1997). The neural correlates of the noradrenergic modulation of human attention, arousal and learning. *European Journal of Neuroscience, 9*, 589–598.

Coull, J. T., Hodges, J. R., & Sahakian, B. J. (1996). The alpha2 antagonist idazoxan remediates certain attentional and executive dysfunction in patients with dementia of frontal type. *Psychopharmacology, 123*, 239–249.

Creelman, C. D. (1962). Human discrimination of auditory duration. *Journal of the Acoustical Society of America, 34*, 582–593.

Crow, T. J., Deakin, J. F. W., & Longden, A. (1977). the nucleus accumbens—possible site of antipsychotic action of neuroleptic drugs? *Psychological Medicine, 7*, 213–221.
File, S. E. (1992). Benzodiazepines and memory. *Clinical Neuropharmacology, 15*, Suppl. 1A, 331A.
File, S. E., Skelly, A. M., & Girdler, N. M. (1992). Midazolam-induced retrieval impairments revealed by the use of flumenazil: a study in surgical dental patients. *Journal of Psychopharmacology, 6*, 81–87.
Fortin, C., & Rousseau, R. (1987). Time estimation as an index of processing demand in memory search. *Perception and Psychophysics, 42*, 377–382.
Fortin, C., & Breton, R. (1995). Temporal interval production and processing in working memory. *Perception and Psychophysics, 57*, 203–215.
Fortin, C., Rousseau, R., Bourque, P., & Kirouac, E. (1993). Time estimation an concurrent nontemporal processing: Specific interference from short-term-memory demands. *Perception and Psychophysics, 53*, 536–548.
Fraisse, P. (1984). Perception and estimation of time. *Annual Review of Psychology, 35*, 1–36.
Frankenhaeuser, M. (1959). *Estimation of time.* Uppsala: Almqvist and Wiksells.
Gerlach, J., & Casey, D. E. (1990). Remoxipride, a new selective D2 antagonist, and haloperidol in cebus monkeys. *Progress in Neuropsychopharmacology and Biological Psychiatry, 14*, 103–112.
Getty, D. J. (1975). Discrimination of short temporal intervals: a comparison of two models. *Perception and Psychophysics, 18*, 1–8.
Gibbon, J. (1977). Scalar expectancy theory and Weber's Law in animal timing. *Psychological Review, 84*, 279–325.
Gibbon, J., & Church, R. M. (1984). Sources of variance in an information processing theory of timing. In H. L. Roitblat, T. G. Bever, & H. S. Terrace (Eds.), *Animal cognition* (pp. 465–488). Hillsdale, NJ: Erlbaum.
Goldman-Rakic, P. S. (1996). Regional and cellular fractionation of working memory. *Proceedings of the National Academy of Sciences of the United States of America, 93*, 13473–13480.
Goldstone, S., & Lhamon, W. T. (1976). The effects of haloperidol upon temporal information processing by patients with Tourette's syndrome. *Psychopharmacology, 50*, 7–10.
Goldstone, S., Nurnberg, H. G., & Lhamon, W. T. (1979). Effects of trifluoperazine, chlorpromazine and haloperidol upon temporal information processing by schizophrenic patients. *Psychopharmacology, 65*, 119–124.
Grondin, S., & Macar, F. (1992). Dividing attention between temporal and nontemporal tasks : A performance operating characteristic -POC- analysis. In F. Macar, V. Pouthas, & W. Friedman (Eds.), *Time, action and cognition : Towards bridging the gap* (pp. 119–128). Dordrecht, The Netherlands : Kluwer Academic Publishers.
Grondin, S., Meilleur-Wells, G., & Lachance, R. (1999). When to start explicit counting in time-intervals discrimination task : A critical point in the timing process of humans. *Journal of Experimental Psychology: Human Perception and Performance, 25*, 993–1004.
Hellström, Å., & Rammsayer, T. (2002). Mechanisms behind discrimination of short and long auditory durations. In J. A. da Silva, E. H. Matsushima, & N. P. Ribeiro-Filho (Eds.), *Annual Meeting of the International Society for Psychophysics* (pp. 110–115). Rio de Janeiro, Brazil: The International Society for Psychophysics.
Hornykiewicz, O. (1972). Neurochemistry of parkinsonism. In A. Lajtha (Ed.), *Handbook of Neurochemistry*, Vol. 7 (pp. 465–501). New York: Plenum.
Janke, W. (1983). *Response variability to psychotropic drugs.* Oxford: Pergamon Press.

Kaernbach, C. (1991). Simple adaptive testing with the weighted up-down method. *Perception and Psychophysics, 49*, 227–229.
Kandel, E. R. (1976). *The cellular basis of behavior*. New York: Freeman.
Killeen, P. R., & Weiss, N. A. (1987). Optimal timing and the Weber function. *Psychological Review, 94*, 455–468.
Kimberg, D. Y., D'Esposito, M., & Farah, M. J. (1997). Effects of bromocriptine on human subjects depend on working memory capacity. *Neuroreport, 8*, 3581–3585.
King, D. J. (1993). Measures of neuroleptic effects on cognition and psychomotor performance in healthy volunteers. In I. Hindmarch & P. D. Stonier (Eds.) *Human Psychopharmacology*, Vol. 4 (pp. 195–209). Chichester: John Wiley.
Köhler, C., Hall, H., Magnusson, O., Lewander, T., & Gustafsson, K. (1990). Biochemical pharmacology of the atypical neuroleptic remoxipride. *Acta Psychiatrica Scandinavica, 82* Suppl., 27–36.
Levitt, H. (1971). Transformed up-down methods in psychoacoustics. *Journal of the Acoustical Society of America, 49*, 467–477.
Luciana, M., Depue, R. A., Arbisi, P., & Leon, A. (1992). Facilitation of working memory in humans by a D2 dopamine receptor agonist. *Journal of Cognitive Neuroscience, 4*, 58–68.
Macar, F., Grondin, S., & Casini, L. (1994). Controlled attention sharing time estimation. *Memory and Cognition, 22*, 673–686.
Maricq, A. V., & Church, R. M. (1983). The differential effects of haloperidol and methamphetamine on time estimation in the rat. *Psychopharmacology, 79*, 10–15.
Mattes, S., & Ulrich, R. (1998). Directed attention prolongs the perceived duration of a brief stimulus. *Perception and Psychophysics, 60*, 1305–1317.
Mattila, M. J., Mattila, M. E., Konno, K., & Saarialho-Kere, U. (1988). Objective and subjective effects of remoxipride, alone and in combination with ethanol or diazepam, on performance in healthy subjects. *Journal of Psychopharmacology, 2*, 138–149.
McClelland, G. R., Cooper, S. M., & Pilgrim, A. J. (1990). A comparison of the central nervous system effects of haloperidol, chlorpromazine and sulpiride in normal volunteers. *British Journal of Clinical Pharmacology, 30*, 795–803.
McGrath, J. E., & Kelly, J. R. (1986). *Time and human interaction*. New York: Guilford Press.
Meck, W. H. (1983). Selective adjustment of the speed of internal clock and memory processes. *Journal of Experimental Psychology: Animal Behavior Processes, 9*, 171–201.
Meck, W. H. (1996). Neuropharmacology of timing and time perception. *Cognitive Brain Research, 3*, 227–242.
Michon, J. A. (1975). Time experience and memory processes. In J. T. Fraser & N. Lawrence (Eds.), *The Study of Time II* (pp. 302–313). New York: Springer.
Michon, J.A. (1985). The compleat time experiencer. In J. A. Michon & J. L. Jackson (Eds.), *Time, Mind, and Behavior* (pp. 21–52). Berlin: Springer.
Mitrani, L., Shekerdjiiski, S., Gourevitch, A., & Yanev, S. (1977). Identification of short time intervals under LSD_{25} and mescaline. *Activas Nervosa Superior, 19*, 103–104.
Müller, U., von Cramon, D. Y., & Pollmann, S. (1998). D1- versus D2-receptor modulation of visuospatial working memory in humans. *Journal of Neuroscience, 18*, 2720–2728.
Münsterberg, H. (1889). *Beiträge zur experimentellen Psychologie*, Heft 2. Freiburg, Germany: Akademische Verlagsbuchhandlung von J. C. B. Mohr.
O'Boyle, D. J., Freeman, J. S., & Cody, F. W. (1996). The accuracy and precision of self-paced, repetitive movements in subjects with Parkinson's disease. *Brain, 119*, 51–70.

Ögren, S.-O., Hall, H., Köhler, C., Magnusson, O., Lindbom, L. O., Ängeby, K., & Florvall, L. (1984). Remoxipride, a new potential antipsychotic compound with selective anti-dopaminergic actions in the brain. *European Journal of Pharmacology, 102*, 459–474.

Ornstein, R. E. (1969). *On the experience of time*. Harmondsworth: Penguin Books.

Park, S., & Holzman, P. S. (1993). Association of working memory deficit and eye tracking dysfunction in schizophrenia. *Schizophrenia Research, 11*, 55–61.

Rammsayer, T. (1989a). Dopaminergic and serotoninergic influence on duration discrimination and vigilance. *Pharmacopsychiatry, 22*, Suppl.1, 39–43.

Rammsayer, T. (1989b). Is there a common dopaminergic basis of time perception and reaction time? *Neuropsychobiology, 21*, 37–42.

Rammsayer, T. H. (1992). An experimental comparison of the weighted up-down method and the transformed up-down method. *Bulletin of the Psychonomic Society, 30*, 425–427.

Rammsayer, T. H. (1993). On dopaminergic modulation of temporal information processing. *Biological Psychology, 36*, 209–222.

Rammsayer, T. H. (1994). Temporal information processing and memory. In L. M. Ward (Ed.) *Fechner Day 94. Proceedings of the Tenth Annual Meeting of the International Society for Psychophysics* (pp. 48–53). Vancouver, Canada: The International Society for Psychophysics.

Rammsayer, T. (1996). Experimental evidence for different timing mechanisms underlying temporal discrimination. In S. C. Masin (Ed.), *Fechner Day 96. Proceedings of the Twelth Annual Meeting of the International Society for Psychophysics* (pp. 63–68). Padua, Italy: The International Society for Psychophysics.

Rammsayer, T. (1997a). Are there dissociable roles of the mesostriatal and mesolimbocortical dopamine systems on temporal information processing in humans? *Neuropsychobiology, 35*, 36–45.

Rammsayer, T. H. (1997b). Effects of body core temperature and brain dopamine activity on timing processes in humans. *Biological Psychology, 46*, 169–192.

Rammsayer, T. H. (1999). Neuropharmacological evidence for different timing mechanisms in humans. *Quarterly Journal of Experimental Psychology, Section B: Comparative and Physiological Psychology, 52*, 273–286.

Rammsayer, T., & Classen, W. (1997). Impaired temporal discrimination in Parkinson's disease: temporal processing of brief durations as an indicator of degeneration of dopaminergic neurons in the basal ganglia. *International Journal of Neuroscience, 91*, 45–55.

Rammsayer, T. H., & Lima, S. D. (1991). Duration discrimination of filled and empty auditory intervals: cognitive and perceptual factors. *Perception and Psychophysics, 50*, 565–574.

Rammsayer, T., & Ulrich, R. (2001). Counting models of temporal discrimination. *Psychonomic Bulletin and Review, 8*, 270–277.

Rammsayer, T. H., Groh, D., & Rodewald, S. (1998). Behavioral effects of pharmacologically induced modulation of excitatory glutamatergic neurotransmission in the brain. In S. Grondin & Y. Lacouture (Eds.), *Fechner Day 98. Proceedings of the Fourteenth Annual Meeting of the International Society for Psychophysics* (pp. 326–331). Quebec, Canada: The International Society for Psychophysics.

Rammsayer, T. H., Hennig, J., Haag, A., & Lange, N. (2001). Effects of noradrenergic activity on temporal information processing in humans. *Quaterly Journal of Experimental Psychology, Section B: Comparative and Physiological Psychology. 54B*, 247–258.

Rammsayer, T. H., Rodewald, S. & Groh, D. (2000). Dopamine-antagonistic, anticholinergic, and GABAergic effects on declarative and procedural memory functions. *Cognitive Brain Research, 9*, 61–71.

Robbins, T. W. (1984). Cortical noradrenaline, attention and arousal. *Psychological Medicine, 14*, 13–21.

Roberts, D. C. S., Price, M. T. C., & Fibiger, H. C. (1976). The dorsal tegmental noradrenergic projection: an analysis of its role in maze learning. *Journal of Comparative Physiology and Psychology, 90*, 363–372.

Russel, R. W. (1987). Drugs as tools for research in neuropsychobiology: A historical perspective. *Neuropsychobiology, 18*, 134–143.

Sawaguchi, T., Matsumura, M., & Kubota, K. (1990). Effects of dopamine antagonists on neuronal activity related to a delayed response task in monkey prefrontal cortex. *Journal of Neurophysiology, 63*, 1401–1412.

Schneider, J. S., & Kovelowski, C. J. (1990). Chronic exposure to low doses of MPTP. I. Cognitive deficits in motor asymptomatic monkeys. *Brain Research, 519*, 122–128.

Seeman, P. (1987). Dopamine receptors and the dopamine hypothesis of schizophrenia. *Synapse, 1*, 133–152.

Seeman, P. (1990). Atypical neuroleptics: role of multiple receptors, endogenous dopamine, and receptor linkage. *Acta Psychiatrica Scandinavica, 82*, Suppl., 14–20.

Solomon, P. R. (1986). Strategies for studying brain-behavior relationships. *Behavioral and Brain Sciences, 9*, 344–345.

Squitieri, G., Cervone, A., & Agnoli, A. (1977). A study on short-term memory in man. Interactions with nooanaleptic and nootropic drugs. In F. Antonelli (Ed.), *Proceedings of the 3rd Congress of the International College of Psychosomatic Medicine* (pp. 742–751). Rome: Pozzi.

Staddon, J. E. R., & Higa, J. J. (1999). Time and memory: towards a pacemaker-free theory of interval timing. *Journal of the Experimental Analysis of Behavior, 71*, 215–251.

Sturt, M. (1925). *The psychology of time*. London: Kegan Paul.

Subhan, Z., & Hindmarch, I. (1983). The effects of midazolam in conjunction with alcohol on iconic memory and free-recall. *Neuropsychobiology, 9*, 230–234.

Thomas, E. A. C., & Weaver, W. B. (1975). Cognitive processing and time perception. *Perception and Psychophysics, 17*, 363–367.

Thompson, R. F. (1976). The search for the engram. *American Psychologist, 31*, 209–227.

Treisman, M. (1963). Temporal discrimination and the indifference interval: implications for a model of the "internal clock". *Psychological Monographs, 77*, 1–31.

Treisman, M., Faulkner, A., Naish, P. L. N., & Brogan, D. (1990). The internal clock: evidence for a temporal oscillator underlying time perception with some estimates of its characteristic frequency. *Perception, 19*, 705–743.

Zakay, D. (1998). Attention allocation policy influences prospective timing. *Psychonomic Bulletin and Review, 5*, 114–118.

Zakay, D. (2000). Gating or switching? Gating is a better model of prospective timing (a response to "switching or gating?" by Lejeune). *Behavioural Processes, 50*, 1–7.

Zakay, D., & Block, R. A. (1996). The role of attention in time estimation processes. In M. A. Pastor & J. Artieda (Eds.), *Time, internal clocks and movement* (pp. 143–164). Amsterdam: Elsevier.

Zakay, D., & Block, R. A. (1997). Temporal cognition. *Current Directions in Psychological Science, 6*, 12–16.

Zakay, D., Block, R. A., & Tsal, Y. (1999). Prospective duration estimation and performance. In D. Gopher & A. Koriat (Eds.), *Attention and Performance* Vol. XVII (pp.557–580). Cambridge, MA: MIT Press.

Part II:

Time as a constituent of information processing

Chapter 7:
Temporal characteristics of auditory event-synthesis: Electrophysiological studies

István Czigler, István Winkler, Elyse Sussmann,
Hirooki Yabe, and János Horváth

Abstract

Mismatch-negativity component of event-related potentials (MMN) is a sensitive indicator of automatic change detection in the auditory modality. This wave is elicited whenever auditory events do not fit the rules stored in sensory memory. No attention to the irregular event is needed to elicit MMN (see Näätänen, 1992, for a review). MMN studies will be reviewed showing that the stream of auditory input is integrated into chunks of approximately 200 ms. However, shorter stimuli within such chunks may preserve their identities, i.e., they are available for further processing.

Temporal integration in the auditory modality

In this paper we suggest that the system of auditory perception is constructed of units of approximately 200 ms duration. The content of these units is preserved in a sensory memory, and used as an implicit representation of the regularities of the auditory environment.

Temporal integration is an essential factor of the acuity of the auditory system. Temporal acuity, however, is markedly different in various domains of auditory perception. To give some examples, in case of spatial localization, temporal resolution is less then 1 ms, and gap-detection threshold is in the range of 1–8 ms (e.g., Bertoli, Heimberg, Smurzynsky, & Probst, 2001, for a review see Moore, 1997). Voice onset asynchrony of 30–40 ms is a cue for phonemic categorization (e.g., Goldstein, 1984), duration/intensity integration is in the 100–200 ms range (Moore, 1997). Duration discrimination performance is better with markers shorter than 200 ms than with longer markers (Rammsayer & Leutner, 1996). Finally, within 200 ms, auditory stimuli are perceived as belonging to the same event (auditory event-synthesis.)

Automatic change-detection in the auditory modality

In this paper we analyze the temporal characteristics of auditory event-synthesis by using a psychophysiological method, the analysis of the mismatch negativity component (MMN) of the event-related brain potentials (see Näätänen, 1992, for a review).

Mismatch negativity is elicited to randomly presented infrequent tones mixed with frequent ones. The infrequent stimuli elicit a negative wave, generated in the auditory cortex (supratemporal regions). MMN is identified as the difference between the event-related potentials (ERPs) to the infrequent (deviant) and frequent (standard) stimuli, maximal over the anterior locations, and peaks between 100 and 200 ms after the onset of a deviant stimulus characteristic. MMN inverts polarity at scalp locations below the Sylvian fissure, i.e., the wave is positive at mastoid electrodes. MMN can be elicited even in the absence of attention to the tones (in typical MMN studies the participants read books they are interested in, play video games or perform demanding tasks). Accordingly, change detection reflected by the MMN is an automatic process. Figure 1 shows a typical MMN record.

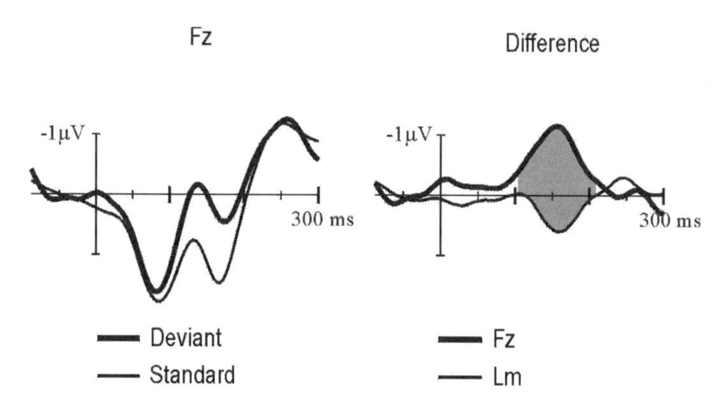

Figure 1: Mismatch negativity (MMN) for frequency change. The frequency of the standard stimuli (p = 0.9) was 700 Hz, the frequency of the deviants (p = 0.1) was 780 Hz. Stimuli have 70 dB intensity and 50 ms duration. Left side: ERPs to the standard and deviant stimuli at midline frontal (Fz) location. Right side: deviant *minus* standard difference potentials at Fz and left mastoid (Lm) locations. Data come from the present authors.

Processes underlying the MMN are explained by the trace-mismatch theory, developed by Risto Näätänen (see Picton, Alain, Otten, Ritter, & Achim, 2000 and Näätänen & Winkler, 1999 and for reviews). According to this theory, MMN is elicited whenever the memory trace acquired by the standard is violated by the characteristics of the deviant stimulus. This way, pre-attentively represented rules can be inferred from the emergence of MMN. The regularities which have been shown to be detected during the automatic (pre-attentive) phase of auditory information processing range from simple repetition rules (parameters of the frequent stimuli) to regularities of complex sequential or spectro-temporal sound patterns (e.g., repetition within a series of two alternating sounds; see Horváth, Czigler, Sussman, & Winkler, 2001).

Temporal characteristics of temporal integration as revealed by the mismatch negativity component (MMN) of event-related potentials

In order to investigate the temporal characteristics of sensory integration, it is supposed that if a stimulus violates the stored regularities of an auditory event within the period of integration, only one MMN will be elicited. However, beyond a critical duration two MMNs will be elicited if a stimulus violates two rules. In order to illustrate the effects of critical duration, we (Winkler, Czigler, Jaramillo, Paavilainen, & Näätänen, 1998) presented auditory stimuli consisting of two segments. Each stimulus started with a segment of constant frequency, followed by a sequence of frequency jump (glide). In separate blocks the deviant differed from the standard either in stimulus intensity (a deviant feature which could be detected at stimulus onset), or in the direction of the frequency jump (ascending *versus* descending), or in both features (double deviant). Note, that the glide deviant could have been detected only after the steady sequence. In the *short separation* condition the steady sequence was 150 ms, while in the *long separation* condition it was 250 ms. As Figure 2 shows, in the *short separation* condition both the intensity deviant and the glide deviant elicited MMN, and the *double deviant* elicited only one MMN. Latency of the MMN was identical to that of the intensity deviant. However, in the *long duration* condition the double deviant elicited two consecutive MMNs (Figure 3). As these results shows, the duration of temporal event-synthesis (temporal window of integration) was approximately 200 ms.

One may say, that the difference between the short and long separation conditions is due to a trivial effect, refractoriness of the MMN generator. This is not the case, however. In a subsequent study (Winkler & Czigler, 1998) two rules of different quality were represented by a single stimulus. Stimuli alternated in frequency, and the standard stimuli had 250 ms duration. Within the sequence, stimuli with the same frequency were repeated infrequently (repetition deviant), furthermore infrequently tones of shorter duration (150 ms) were delivered (duration deviant), and finally, stimuli with repeated frequency had also shorter duration (double deviant). In spite of the 150 ms temporal separation, double deviants elicited two MMNs.

Analyzing the magnetic equivalent of MMN (mMMN), Yabe et al. (1998) developed a method for the quantitative assessment of the duration of the temporal window of integration. In sequences of identical tones the stimulus onset asynchrony (SOA) was varied between 150 and 350 ms. Tones were sometimes omitted from the sequences. Unlike at longer SOA, when the SOA was shorter than 200 ms, stimulus omission elicited a negative wave, similar to the MMN component. In the "standard" windows of integration (i.e., without stimulus omission) there were two tones. In the "deviant" windows of integration (i.e., with stimulus omission) there was only one. The negative wave is a difference between the ERP to the standard and deviant windows of integration. Therefore the negativity to the stimulus omission seems to be a MMN.

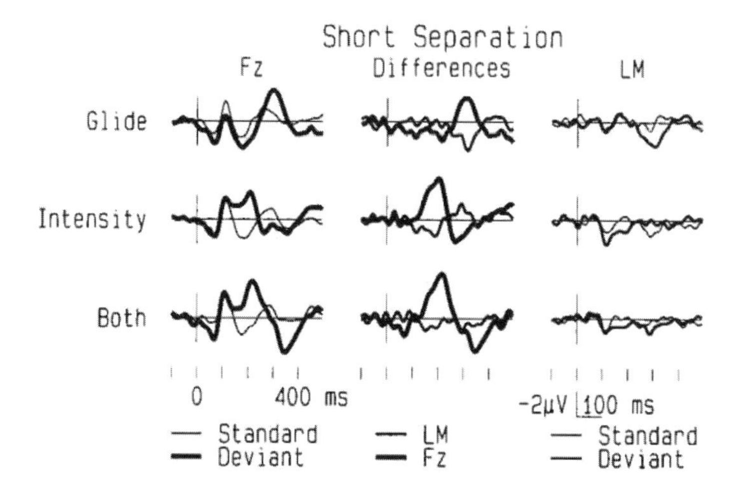

Figure 2: Grand average (n = 10) responses to standard (thin line) and deviant (thick line) responses recorded at Fz (left column) and the left mastoid (right column) in the *short separation* condition (150 ms separation between the onset of the critical elements). The corresponding deviant *minus* standard differences are shown in the central column. Rows in the figure present conditions with different deviations: glide, intensity, and double deviant. Adapted from Winkler, Czigler, Jaramillo et al. (1998, p. 497).[7]

In a subsequent study Yabe, Winkler, Czigler et al. (2001) investigated the content of the temporal window of integration. Tone sequences were constructed from either alternating tones of two similar frequencies (3000 and 2800 Hz) or two highly different frequencies (3000 Hz and 500 Hz). In both cases the SOA was 125 ms. Omitted stimuli in the sequences of similar tones elicited mMMN while there was no such negativity in the sequences of different tones. Stream segregation of the auditory input is a basic perceptual process (Bregman, 1990). As the results of our study show, stream segregation precedes the process of temporal integration. While stimuli in the sequence of similar tones were organized into a single stream, sequence of the highly different tones consisted of two streams. In the former case omitted stimuli constituted deviant temporal windows (instead of two stimuli, only one stimulus per stream within 200 ms), while in the latter case all temporal windows contained only

[7] Reprinted with permission from Lippincott Williams & Wilkins.

one stimulus per stream (note that even without omission the SOA between two stimuli of a particular stream was longer than 200 ms).

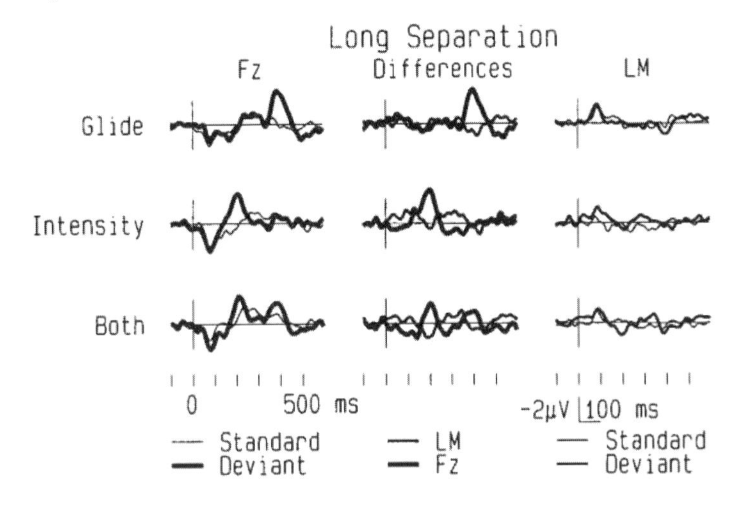

Figure 3: Grand-averaged (n = 10) responses recorded in the long separation condition (250 ms separation between the onsets of the critical stimulus elements. For description see Figure 2. Adapted from Winkler, Czigler, Jaramillo et al. (1998, p. 498).[8]

If stimuli of an auditory stream are organized into units of 200 ms, only the first deviant elicits MMN when two consecutive deviants are presented with SOA less than 200 ms. Figure 4 presents the results of such an experiment, designed after Sussman, Winkler, Ritter, Alho, and Näätänen (1999). Within a sequence of standard tones, pitch deviants were followed by intensity deviants. In various sequences the SOA was varied between 150 and 300 ms. As Figure 4 shows, unlike the first (frequency) deviant, the second (intensity) deviant elicited MMN only if the SOA was at least 200 ms.

The studies so far reviewed indicated a fixed, 200 ms integration period. However, according to a recent study (Sussman & Winkler, 2001), duration of the temporal integration period depends on the *context* of auditory stimulation. Within a sequence of standard tones (SOA = 150 ms) either single deviants or pairs of deviants were presented. At the beginning of these mixed sequences the second member of the deviant pair did not elicit MMN. However, at the end of the sequence the second member of the deviant pair elicited MMN. In the control condition with only paired deviants, the second member of the pair did not elicit MMN at the end of the sequence.

To summarize the results so far reviewed, it seems that the auditory system constructs units of 200 ms duration. This is an automatic, but adaptive process. These

---

[8] Reprinted with permission from Lippincott Williams & Wilkins.

units are stored in the sensory memory, and the content of the units can be used in further processing.

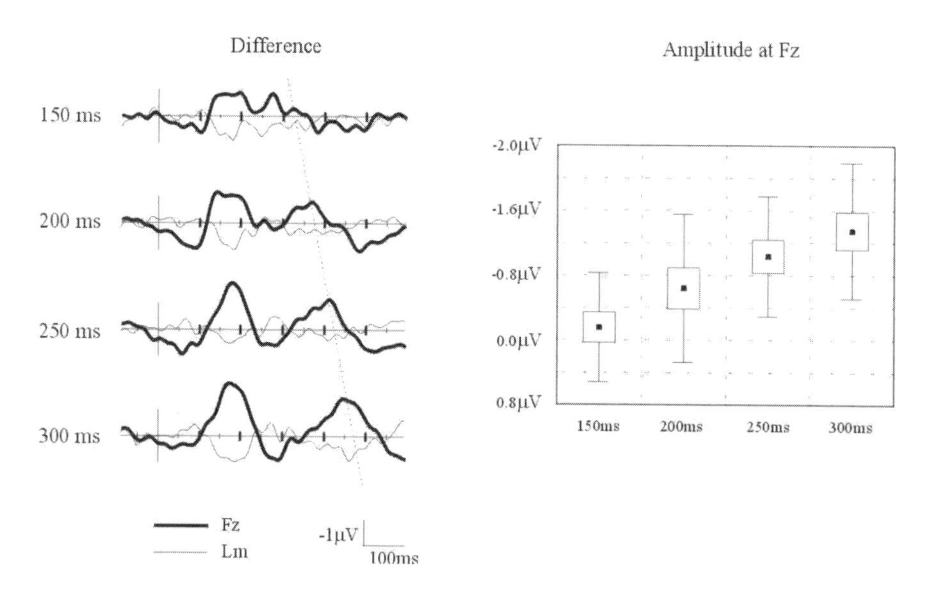

Figure 4: Grand average difference potentials (left side) and the amplitude of the MMN to the second deviant to consecutive deviant stimuli. The SOA was varied between 150 and 300 ms. Note, that no MMN was elicited at 150 ms SOA (Czigler, Horváth, & Winkler, unpublished data).

Duration of the auditory memory underlying the mismatch negativity (MMN) component of event-related potentials

The next question concerns the duration of the store of such units. A direct treatment of this issue is the investigation of the MMN as a function of SOA. Results of such studies (Böttcher-Gandor & Ullsperger, 1992; Czigler, Csibra, & Csontos, 1992) show that MMN is elicited at 4–10 sec SOA, but no MMN is elicited at SOA longer than 11–15 sec (Cowan, Winkler, Teder, & Näätänen, 1993). Due to the short duration of the memory underlying MMN, it is expected that the system is insensitive to regularities at longer time scale. In a well-known study Scherg, Vajsar and Picton (1989) compared the ERPs to infrequent ($p = .2$) stimuli from random and regular (AAAABAAAAB…) sequences with 1300 ms SOA. The infrequent (deviant) stimuli elicited identical MMNs in the two types of sequences. This result can be interpreted as showing that the system underlying MMN is insensitive to complex/abstract rules, and governed by a simple trace strength principle. According to this view, a sensory trace is formed on the basis of the physical features of the standard stimuli. MMN is a reflection of the mismatching physical features. Therefore, emergence of MMN is dependent of the probability structure of the sequence, but independent of the regularity of the sequence. This mechanical view is questioned by results of many studies showing effects of higher order regularities on MMN (e.g., Horváth et al., 2001). Sussman, Ritter, and Vaughan (1998) offered an alternative explanation of the Scherg et al. (1989) results. They repeated the study, together with a short SOA (100 ms) condition. Infrequent stimuli in the random sequences elicited MMN in both SOA conditions. However, unlike in the long SOA condition, in the short SOA condition infrequent stimuli of the regular sequences (AAAAB-AAAAB…) didn't elicit MMN. According to the authors, in the long SOA condition, repetition period of the temporal pattern (AAAAB) was beyond the duration of the memory underlying MMN, i.e. the system was unable to record the regularity. On the other hand, in the short SOA condition the system registered the pattern; thus the appearance of the infrequent stimulus was fully predictable event, and this event did not elicit MMN. At this point it should be emphasized again, that throughout the MMN experiments the participants did not pay attention to the tones. However, presenting a long SOA sequence under auditory attention ("Pay attention to the tones, and try to identify the structure of the sequence!"), an adult easily detects the AAAABAAAAB… regularity. We suggest that one of the roles of attentive processing is to encompass longer ranges in the temporal domain.

Implications for the perception/experience of temporal duration

The concept of internal clock has a central role in many theories of time-related behavior (see the chapters by Church et al., Grondin, Klapproth and Wearden in this volume). However, attention to temporal *versus* non-temporal cues/events and the number of contextual changes is a crucial determinant of time experience (see Block, this volume, chapter 3). Contrary to suggestions of the importance of such cognitive processes, the possible role of pre-attentive (automatic) processes are hardly considered in the modern theories of duration experience. However, such processes may be incorporated into the theories emphasizing the role of contextual change (Block, 1985, and Block, this volume, chapter 3) or into the so-called *structural remembering model* proposed by Boltz (1998). At any rate, potential effects of unattended stimuli, and the structure of unattended sensory input on time estimation is a matter of empirical research.

References

Bertoli, S., Heimberg, S., Smurzynski, J., & Probst, R. (2001). Mismatch negativity and psycho-acoustic measures of gap detection in normally hearing subjects. *Psychophysiology, 38,* 334–342.

Böttcher-Gandor, C., & Ullsperger, P. (1992). Mismatch negativity in event-related potentials to auditory stimuli as a function of varying interstimulus interval. *Psychophysiology, 29,* 546–550.

Block, R. A. (1985). Contextual coding in memory. Studies of remembered duration. In J. Michon & J. Jackson (Eds), *Time, action and behavior* (pp. 169–178). Heidelberg: Springer.

Boltz, M. B. (1998). The process of temporal and nontemporal information in the remembering of event durations and musical structure. *Journal of Experimental Psychology: Human Perception and Performance, 24,* 1087–1104.

Bregman, A. S. (1990). *Auditory scene analysis.* Cambridge MA.: MIT Press.

Cowan, N., Winkler, I., Teder, W., & Näätänen, R. (1993). Memory prerequisites of mismatch negativity in the auditory event-related potentials (ERP). *Journal of Experimental Psychology: Learning, Memory and Cognition, 19,* 919–921.

Czigler, I., Csibra, G., & Csontos, A. (1992). Age and inter-stimulus interval effect on event-related potentials to frequent and infrequent auditory stimuli. *Biological Psychology, 33,* 195–206.

Goldstein, E. B. (1984). *Sensation and perception.* Belmont: Wadsworth Publishing Company.

Horváth, J., Czigler, I., Sussman, E., & Winkler, I. (2001). Simultaneously active pre-attentive representations of local and global rules for sound sequences in the human brain. *Cognitive Brain Research, 12,* 131–144.

Moore, B. C. J. (1997). *An introduction to the psychology of hearing.* San Diego, CA.: Academic Press.

Näätänen, R. (1992). *Attention and brain function.* Hillsdale, NJ: Erlbaum.

Näätänen, R., & Winkler, I. (1999). The concept of auditory stimulus representation in cognitive neuroscience. *Psychological Bulletin, 125,* 826–859.

Picton, T. W., Alain, C., Otten, L., Ritter, W., & Achim, A. (2000). Mismatch negativity: Different water in the same river. *Audiology and Neuro-Otology, 5,* 111–139.

Rammsayer, T. H., & Leutner, D. (1996). Temporal discrimination as a function of marker duration. *Perception and Psychophysics, 58,* 1213–1223.

Scherg, M., Vajsar, J., & Picton, T. W. (1989). A source analysis of the late human auditory evoked potentials. *Journal of Cognitive Neuroscience, 1,* 336–355.

Sussman, E., Ritter, W., & Vaughan, H. G. Jr. (1998). Predictability of stimulus deviance and the mismatch negativity. *NeuroReport, 9,* 4167–4170.

Sussman, E., Winkler, I., Ritter, W., Alho, K., & Näätänen, R. (1999). Temporal integration of auditory stimulus deviance as reflected by mismatch negativity. *Neuroscience Letters, 264,* 161–164.

Sussman, E., & Winkler, I. (2001). Dynamic sensory updating in the auditory system. *Cognitive Brain Research, 12,* 431–439.

Winkler, I., Czigler, I., Jaramillo, M., Paavilainen, P. & Näätänen, R. (1998). Temporal constraints of auditory event synthesis: Evidence from ERPs. *NeuroReport, 9,* 495–499.

Winkler, I., & Czigler, I. (1998). Mismatch negativity: Deviance detection or the maintenance of the 'standard'. *NeuroReport, 9,* 3809–3813.

Yabe, H., Tervaniemi, M., Sinkkonen, J., Huotilainen, M., Ilmoniemi, R. J., & Näätänen, R. (1998). Temporal window of integration of auditory information in the human brain. *Psychophysiology, 35,* 615–619.

Yabe, H., Winkler, I., Czigler, I., Koyama, S., Kakigi, R., Sutoh, T., Hiruma, T., & Kaneko, S. (2001). Organizing sound sequences in the human brain: The interplay of auditory streaming and temporal integration. *Brain Research, 897*, 222–227.

Chapter 8: Exploring the timing of human visual processing*

SIMO VANNI, MICHEL DOJAT, JAN WARNKING, CHRISTOPH SEGEBARTH, and JEAN BULLIER

Abstract

Simulations with neuromagnetic data indicate that activity in the visual cortices is unlikely to be fully disentangled either with dipole or distributed inverse solutions without prior knowledge of sources. Thus it is necessary to combine the spatial resolution of functional magnetic resonance imaging (fMRI) and the temporal resolution of magneto- (MEG) or electroencephalography (EEG) to enhance quantitative studies of distributed visual processing in humans. The combination has proved not to be straightforward, however. The statistical activation maps in fMRI include false positive and negative voxels, and the results are susceptible to motion and other sources of artefacts. On the other hand, EEG and MEG models are sensitive to noise, and a large number of sources render solutions unstable. Eventually, due to different generators and sensitivities of signal in fMRI versus EEG/MEG, it is not likely that direct fMRI-to-source approach would work.

We extracted the position and orientation of a subset of dipoles from fMRI data; only dipoles included in retinotopically mapped regions were taken as "truth." Other fMRI-seeded dipoles were included if they were replicable across subjects, and eventually the unexplained variance in the EEG signal was modelled the traditional way. By carefully planning of the paradigm we were able to have good signal to noise in EEG and a limited number of active regions in fMRI—both prerequisites for reasonable EEG source modelling.

In future, correction of echo-planar imaging distorsions could provide enhanced alignment between anatomical and functional magnetic resonance (MR) images for the optimal position and orientation of the dipoles. In addition, individual finite element conductor model in EEG and MEG would probably bring the solution minima closer to real positions of active cortex. In spite of the limitations, we believe that combination of the methods is a major step forward in quantifying the time and strength of neural response at the resolution of a functional area.

* We thank Chantal Delon-Martin for help in fMRI data acquisition and Andrew James for multifocal mapping of visual field response amplitudes of some of our subjects. Margot Taylor provided us an additional amplifier for the EEG recordings, and Renaud Lestringant helped with coordinate transformations between EEG and MRI data. We thank all subjects, who showed remarkable persistence with participation in measurements at two different cities in at least three different recording sessions. This work was supported by European Commission grant MCFI-2000-01134 from Quality of Life and Management of Living Resources program.

Visual network operations in the human cerebral cortex

When we open our eyes, we have an instantaneous perception of our visual environment. Visual processing takes time, however. The first cortical responses from human primary visual cortex (V1) emerge about 50–55 ms after the onset of a stimulus (Tzelepi, Ioannides, & Pghosyan, 2001; Vanni, Tanskanen, Seppä, Uutela, & Hari, 2001b; Di Russo, Martinez, Sereno, Pitzalis, & Hillyard 2002; Foxe & Simpson, 2002). From V1, activation spreads to other cortical areas and, as is known in the monkey, each functional area presents a typical range of activation onsets (Nowak & Bullier, 1997). In monkeys, the most evident differences in timing have been observed between the so-called dorsal and ventral visual areas (Nowak & Bullier, 1997; Schmolesky et al., 1998; Schroeder, Mehta, & Givre, 1998). Several areas in the occipito-parietal dorsal stream are activated within milliseconds after V1, while for the ventral information stream, the successive activation of sequentially connected areas such as V1, V2 and V4 occurs with a 10 ms delay on the average. In humans, neuromagnetic recordings indicate both parallel and serial activations, with phases or durations of activation varying between different areas, sometimes with clear individual variability (Salmelin, Hari, Lounasmaa, & Sams, 1994; Vanni & Uutela, 2000). The largest activations occur at the onset of stimuli, even when the stimulus is oscillatory (Vanni, Raninen, Näsänen, Tanskanen, & Hyvärinen, 2001a). After pattern reversal with no task, activation disappears in about 400 ms (Vanni et al., 2001b). This time lapse is comparable to the mean fixation time between saccades (about 300 ms). With informative stimuli, fixation may persist significantly longer, with a corresponding extension of the period of visual processing. Extended processing times have been found for example with visual object detection (Vanni, Revonsuo, Saarinen, & Hari, 1996) or naming (Salmelin et al., 1994). Computations in this system are highly efficient. The first responses related to stimulus categorization are detectable in human evoked responses about 150 ms after stimulus onset (Thorpe, Fize, & Marlot, 1996), i.e., not more than 100 ms after the first cortical activation.

Visual processing is of course much more than an activation sequence. This sequence is only a reflection of the underlying neural computations, and the causal relationship between macroscopic activation in a particular functional area and the corresponding computational task is unclear. To clarify the computational tasks at a particular area, unit recordings in animals are important, and to understand the network operations involved, one of the first steps to be undertaken is to identify the timing of the different areas involved. This provides insight into the dynamics of the availability of specific resources, such as large receptive fields or specific sensitivities, as well as into the information flow between the processing nodes.

There is a limit to the homology between species, and clinical applications must necessarily be based on human studies. Because cortical surface in humans is much larger than in monkey, thus inducing larger distances between functional areas, specific computational solutions might ensue. Increase in distance means increase in processing time, provided that fiber properties stay the same. Interestingly, in monkeys the feedforward and feedback connections between different functional areas conduct information much faster than the horizontal connections within an area (Girard, Hupe, & Bullier, 2001). Also, areas high up in the functional hierarchy may strongly alter early processing at the low-level areas (Hupe et al., 1998; Bullier, 2001; Hupe et al., 2001). It is thus of considerable interest to find out the sequence of activation of the different areas in humans.

Earlier related attempts have been limited to separating activation of the primary visual cortex V1 from activations from extrastriate areas, without specifically trying to identify latencies of activations. Now that a number of visual areas may be delineated in the human individual, based on retinotopic mapping and on other activation studies with functional magnetic resonance imaging (fMRI), this information may be used to constrain the solutions of the electromagnetic inverse problem and hence to measure latencies of activation at specific visual cortical areas in the human. This chapter describes our first attempts into this direction. What we have learned from these first experiments raises hope that fMRI may indeed be applied to constrain the solutions of both electroencephalography (EEG) and magnetoencephalography (MEG) source analysis. Careful planning of the paradigm, familiarity with the data, and a good understanding of the limits of each imaging modality are central.

Imaging the dynamic network

Why do we need to combine fMRI with EEG and MEG?

This chapter introduces the methodological approach we have used, with special emphasis on the limitations and quality controls needed. Combining the high spatial resolution of fMRI with the high temporal resolution of EEG and MEG has become a major issue in human brain imaging. MEG presents the advantage with respect to EEG that the scalp and the skull are transparent to magnetic fields. However, even with MEG, the ambiguity of the inverse problem, i.e., that there exists no mathematically unique source configuration in the brain for a given signal distribution outside the head (von Helmholtz, 1853; Hämäläinen, Hari, Ilmoniemi, Knuutila, & Lounasmaa, 1993), essentially renders all simultaneous sources within a distance of about 3 cm difficult to separate (Uutela, Hämäläinen, & Somersalo, 1999). Especially for the visual system, where even simple stimuli evoke complex networks of parallel processing, simulations show that whole-head multi-channel MEG recordings do not allow to correctly detect the dynamics at distinct functional areas without introducing a priori spatial information (Stenbacka, Vanni, Uutela, & Hari, 2002). fMRI is thus necessary to uniquely localize active areas. fMRI tells us where activation appears for a particular paradigm. In addition, a subset of the human visual areas can be delineated with specific fMRI mapping protocols. Such studies exploit the retinotopic properties of low-order visual areas, such as V1, V2, V3/Vp and V3a/V4v, or the sensitivity of certain areas to specific features of visual stimuli, such as motion in the case of area V5.

The major steps of our approach are displayed in Figure 1, and extensively discussed elsewhere (Vanni et al., submitted-b). Source analysis in EEG and MEG critically depends on signal-to-noise ratio, and thus planning of the paradigm must take this into account. Because some individuals show weak evoked EEG responses and others strong sources of biological noise (such as abundant spontaneous oscillations), we first scanned our subjects to retain only those who exhibited decent evoked responses (two subjects out of nine were excluded).

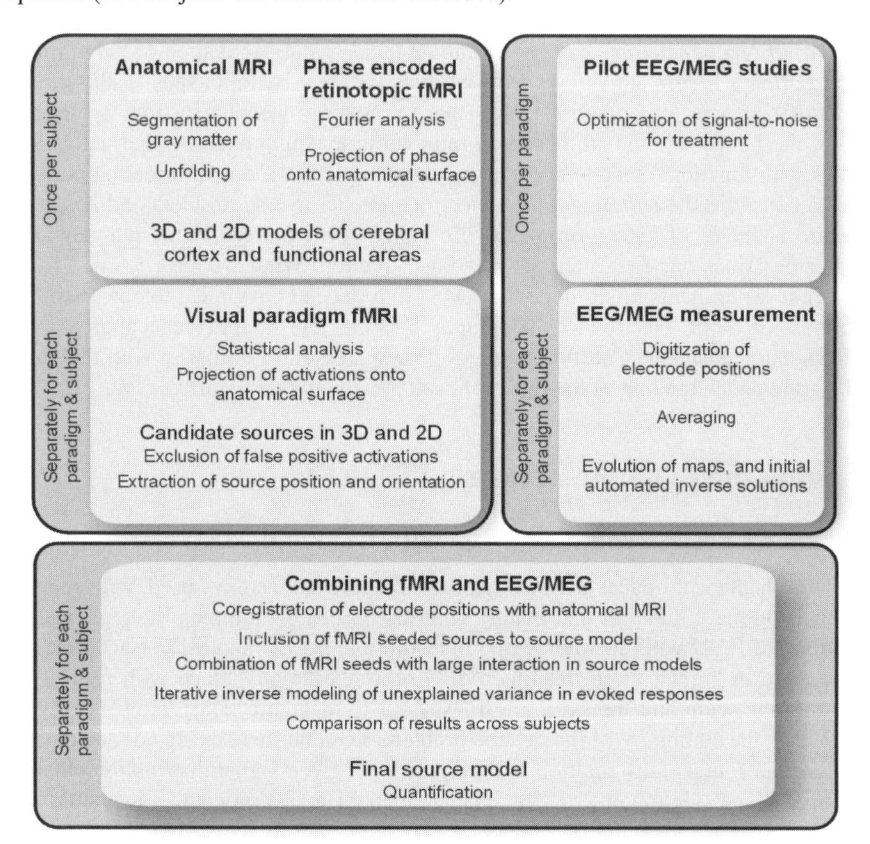

Figure 1: The process of assisting EEG/MEG source modeling with seeds from fMRI can be divided into three sections: fMRI measurements and analysis (left upper box), evoked response measurements and analysis (right upper box), and combination analysis (bottom box).

Traditionally, to improve signal-to-noise ratio for EEG source modeling, the data are averaged across subjects before inverse modeling. Such averaging has been abandoned by the MEG community because of the strong inter-individual variability. Due to the large variability in individual cortical anatomy, and therefore in the phases and polarities of evoked responses in single channels, averaging the channel responses across subjects results in spatial and temporal smoothing. It therefore hampers separability of responses from different functional areas. To be able to separate the timing in discrete areas, we performed all source analyses on individual data. High signal-to-noise level was achieved by averaging a large number of responses (280) per category. The same individuals were scanned with fMRI in order to map the retinotopic areas, and then the visual paradigm from the EEG experiment was carefully replicated in the MRI scanner. The resulting activations were projected onto the anatomical surface reconstruction of the occipital cortex, and the positions and orientations of the active regions were seeded into the multidipole EEG source model. Unexplained variance was modeled with additional sources, thus resulting in a combined model, some of the sources having been obtained from fMRI and others from a traditional inverse localization algorithm.

Planning the paradigm

Identifying the sources of EEG/MEG signals requires a limited number and, if possible, a limited extent of active brain areas. Complex electrical activation in the brain renders the inverse solutions unstable and reduces spatial accuracy. Limited extent is beneficial because wide activation area is difficult to model with one dipolar source. Thus, we chose to use relatively small peripheral pattern stimuli in a passive paradigm. Furthermore, we chose to place the stimuli in the inferior rather than in the superior visual hemi-field, since the former is known to produce stronger signals in MEG and EEG (Portin, Vanni, Virsu, & Hari, 1999; James, 2003). In addition to a simple activation pattern, information about the possible source configurations from prior MEG/EEG data greatly reduces the ambiguity in source localization.

Data in this chapter were acquired with checkerboard pattern onset stimuli. One or two patterns appeared from gray background and stayed stable on the display for 500 ms, with 1 s inter-stimulus interval. The paradigm and results from these experiments are detailed elsewhere (Vanni et al., submitted-a).

Acquiring and understanding the fMRI seeds

We chose to use direct fMRI seeding. In this approach, the same paradigm was replicated in EEG and in fMRI, and the fMRI activations were used to position and orient the EEG dipoles. Figure 2 shows, as an example, a dipole positioned at an fMRI activation in area V2, in one of our subjects. Only those activations which were within the first four retinotopic areas (V1, V2, V3, V3A) were used as seeds, for which the position agreed with the retinotopic representation of the stimulus. This was considered mandatory because a variable number of obviously false positive voxels were obtained in the statistical maps of the occipital cortex. Examples of such presumably false positive activations are present at the bottom and at the top of the flat map in Figure 2. Also, some subjects showed little activation in the retinotopic areas. Therefore, we contrasted for those subjects all conditions with a particular stimulus to all corresponding control conditions to increase statistical sensitivity (Price & Friston, 1997), and to detect some activity in all subjects. Thus, retinotopic mapping turned out to be a very important intermediary step to reject both false positive and false negative fMRI findings. It remains to be understood where the variability in fMRI results comes from.

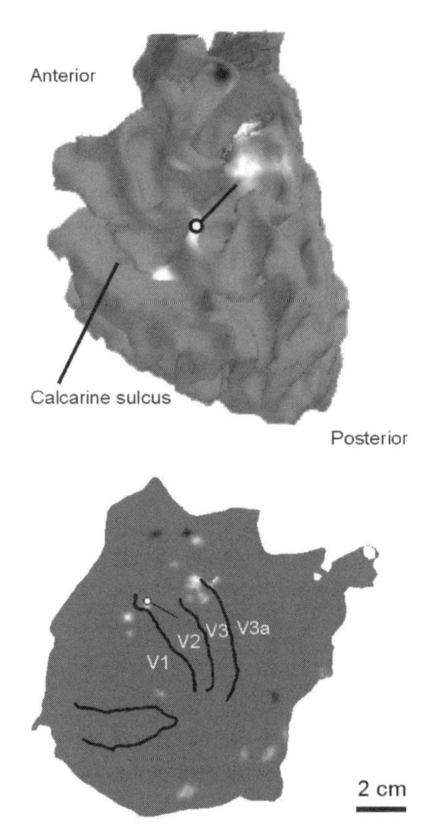

Figure 2: Three-dimensional model of the right occipital lobe of one subject. Activation in fMRI has been projected onto this model, and a dipole at V2 is shown both on the three- and two-dimensional views of the cortex.

A difficulty may lie in the different origins of the signals detected with the different methods. While the EEG and MEG signals originate directly from the synaptic activity (Hämäläinen et al., 1993), blood oxygenation level dependent (BOLD) MRI signal has a cerebrovascular origin and is only an indirect reflection of neuronal activity. In addition to the difference in the link between neural activity and the observed signal, EEG/MEG on the one hand and BOLD fMRI on the other hand are affected by distinct sources of noise and different types of errors arising from signal analysis.

The origin of fMRI BOLD signal is extensively reviewed elsewhere (see Chen & Ogawa, 1999), and only issues relevant to the seeding procedure are discussed here. Deoxyhemoglobin acts as an endogenous contrast agent for the BOLD signal by perturbing the static magnetic field on a microscopic scale. The amount of deoxyhemoglobin in capillary and venous blood depends locally on oxygen consumption, which is linked to local metabolic needs and to local oxygen supply which is further linked to local blood flow. Positive BOLD signal is generated when blood flow increase exceeds what would be necessary to cover the increase in oxygen consumption so that the concentration of deoxyhemoglobin locally decreases. These changes vary from one brain area to the other, as well as between individuals. Thus such direct comparisons of the BOLD signal are not meaningful.

When the anatomical and functional MR images are coregistered, there may be distortions of the functional data with respect to the anatomical data, which may be difficult to recover. This leads to misregistration of the functional responses. This is especially critical in surface oriented analysis where the signal may erroneously be assigned to the wrong bank of a sulcus or a gyrus, thus ending up strongly misregistered on a map of brain areas. Furthermore, fMRI images are subjected to several preprocessing steps, including motion correction, and eventually the significance of BOLD signal change is estimated by statistical models. Typically, a model of signal change (i.e., anticipated BOLD time course) is tested against the measured MR signal time course. Even within individuals, with the same paradigm replicated several times, the statistical maps show significant variance of unknown origin (McGonigle et al., 2000). Whether this variance reflects changes in the measurement conditions or changes in the physiological response is unclear. Moreover, BOLD responses may be detected at relatively large distances from activated cortex, due to macrovascular effects (Disbrow, Slutsky, Roberts, & Krubitzer, 2000). All these factors taken together, it is important to realize that fMRI presents a number of limitations and weaknesses and that it is not giving the "whole truth and nothing but the truth" of brain activation. One session with a single subject might not provide a very reliable reflection of the physiological responses investigated. On the other hand, single subject analysis is needed for EEG/MEG, due to the large variability in the individual brain anatomy (Amunts, Malikovic, Mohlberg, Schormann, & Zilles, 2000) causing inevitable variability in signal distribution in a given channel, or shifts in latencies within or between individuals (Kaneoke, Bundou, & Kakigi, 1998; Vanni & Uutela, 2000).

It is unclear to what extent the different time scales of the phenomena measured with electro-magnetic and with metabolic functional imaging techniques might be the source of incompatibility when combining brain activities measured with these different methods. It is likely that quantitative maps of neural activity will differ for different methods. On the one hand, increases in neural synchrony, to which EEG and MEG are particularly sensitive (Hari, 1990), might produce little changes, if any, in firing rate of ongoing activity and thus induce little changes in metabolic demands and in hemodynamic measures. This might be a problem at the main generators of 10–20-Hz activity close to primary sensory areas (Hari & Salmelin, 1997), where changes often appear simultaneously with stimulus onset. On the other hand, asynchronous start of evoked activity is certainly easier to detect with fMRI than with EEG/MEG, and this should be taken into account when planning the paradigm. In theory, long duration of neural activation (several seconds or more) should also be more easily detectable with fMRI, but it is not always easy to provoke a prolonged neural activation. For example stable stimuli might not produce much neural activation in the first place.

Despite the different time scales and the fact that the BOLD signal is only an indirect marker of neural activity, the two types of signals are undoubtedly linked. In monkeys, the strongest correlations to BOLD signals can be found with local field potentials reflecting synaptic activity rather than with multi-unit response reflecting action potentials (Logothetis, Pauls, Augath, Tinath, & Oeltermann, 2001). In human V1, for several stimulus types, fractional changes in blood flow and oxygen consumption were found to be linearly coupled with a ratio of approximately 2:1 (Hoge et al., 1999), and within a recording session and area, the BOLD signal may well be proportional to the average neural activity (Boynton, Engel, Glover, & Heeger, 1996).

Applying fMRI seeds to a source model

Dipoles are common models for electromagnetic inverse solutions (Lütkenhöner et al., 2000). They model brain activity with a point-like source with either fixed or freely rotating orientation. If we were able to generate only one significantly active brain region at a time, there would be little problem in localizing activity with good temporal resolution with either EEG or MEG alone. However, in real data this is never the case. Both with pattern onset and pattern reversal visual stimuli, even early cortical activation appears to be a combination of signals from different areas (Tzelepi et al., 2001; Vanni et al., 2001b). It is important to understand that two (or more) neighboring EEG or MEG sources with simultaneous activity cannot be correctly identified without a priori knowledge about the source positions and orientations, although wide angle between the two sources enhances separability in inverse modeling algorithms (Mosher, Spencer, Leahy, & Lewis, 1993; Lütkenhöner, 1998; Uutela et al., 1999). In their sophisticated simulation study, Liu et al. (1998) showed the importance of a priori orientation information in reducing crosstalk between sources.

Due to the possibility of false positive and false negative regions of activation in fMRI, and due to the sensitivity of EEG/MEG source models to false or missing dipoles, one should avoid fully constraining the source model by fMRI data (Liu, Belliveau, & Dale, 1998; Dale et al., 2000). In our approach, we accepted seeding only those areas which we a priori anticipate to be active, i.e., the retinotopic areas.

Even with a priori knowledge about the position and orientation of dipolar sources derived from fMRI, the task of addressing EEG or MEG signals correctly to distinct brain areas is not easy. Given a dipole modeling an active brain area, crosstalk from activation in another area may interfere with the dipole amplitude. Especially, when the other active area has not been included in the dipole model with a source of its own, a large part of its activity may be projected onto the existing dipole (Liu et al., 1998). Crosstalk may emerge also between two existing dipoles. In this case, all activation that is not fully aligned with one dipole (extended source, orientation error of the dipole, or noise) leaves a residual, free to interfere with the other dipole. Crosstalk may be suspected if neighboring sources show similar source waveforms (Scherg & Berg, 1991). In our study, if we suspected significant crosstalk between sources, they were averaged together and only one source was used in the model. If crosstalk appears to be present, but the sources are not fully aligned, it may be better to leave both sources in the model, but report the average of the two source waveforms. This way one does not unnecessarily increase unexplained variance in the model, which might affect yet another source. On the other hand, one can rely upon different signals in different sources; if the seeds are at correct positions, and source waveforms show different phases of the deflections, it cannot be due to both sources measuring only the same signal. Obviously, a large number of EEG channels help in telling apart nearby sources (Srinivasan, Tucker, & Murias, 1998). In MEG this is not so much an issue since most results today are measured with a fixed and often quite large number of channels.

One source of error, which was initially a problem in one of our seven subjects, is the misalignment of the digitized electrode positions in the anatomical MRI coordinate system. In our case, misalignment resulted in translation of the seeded dipoles, and this was corrected manually by translation of the dipole positions until the V1 dipole coincided the V1 activation in the brain.

When analyzing the evoked potentials with BESA software (Megis Software Inc.), rather than obtaining beautiful activation patterns at the retinotopic areas after having included the seeded sources, usually noise and crosstalk were predominant. Therefore, noise was modeled next with one or two signal space vectors (Tesche et al., 1995; Uusitalo & Ilmoniemi, 1997) placed at pre-response interval (–100 to +50 ms from stimulus onset), which clearly increased separability between sources at the retinotopic areas. Outside the retinotopic areas, unexplained variance was modeled with additional sources. The left and right lateral occipital regions, sometimes close to the temporo-occipital border, were active in all subjects. We describe this region with a name LOV5, since it was unclear which functional area (the lateral occipital complex or the visual motion sensitive area V5, or some other area close by) corresponds to this source. In addition, some other sources were more sporadically active, but for them no group level statistics was possible, because they did not represent any obvious single functional area.

Sometimes similar patterns of activation emerged in succession from V1 to V3, area V3a showing relatively short-latency onset but prolonged early activation (Figure 3), and sometimes the retinotopic areas showed very similar waveforms. A closer look revealed that the sources producing similar waveforms in similar phases had only a narrow angle between them in 3D, suggesting that our 64-ch EEG just did not have enough spatial resolution to tell these signals apart. In contrast, in some cases, such as the one presented in Figure 3, whereby the signals have distinct phases, the active areas in V1, V2 and V3 were almost spatially orthogonal to each other.

After modeling the activity with 5 to 9 sources (including the retinotopic seeded sources), source waveforms of corresponding brain regions were averaged across subjects and quantified.

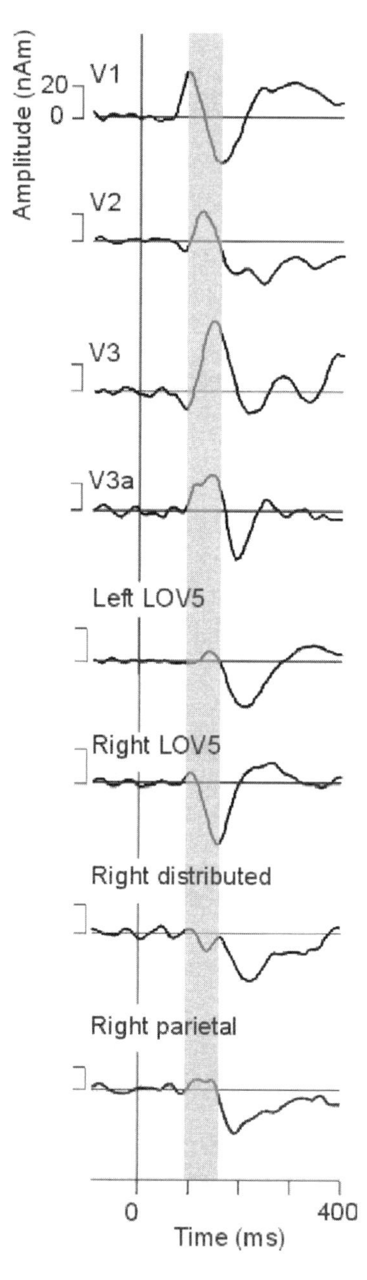

Figure 3: Source waveforms of a multidipole model of a single subject. The first four sources are directly seeded from fMRI whereas the rest have been inverse modeled. The gray area indicates time from 95 to 160 ms.

Dynamics of pattern onset processing in humans

Figure 4 shows the group mean source waveforms for a stimulus in the left inferior visual quadrant. V1+ includes all V1 sources. The plus sign indicates that this source includes some V2 sources when it was impossible to separate them from the V1 sources, and perhaps other activation nearby not included in the model, but which may contribute to the waveform. The V3V3a+ includes all V3 and V3a sources and some V2 sources when they could not be separated. The dipoles at V1 and at V3/V3a were almost always different enough to be separated. The first peak in V1+ had a latency of 83 ms and the first negative peak in the V3V3a+ waveform showed the same latency. The phase similarity was checked and found in the individual data, too. Thus it could be due either to leakage from V1, or to activation of a third nearby source, not included in the model. The second peak (139 versus 127 ms in the V1+ and V3V3a+ respectively) and the third peaks were not simultaneous. The right and left LOV5 were found in all subjects, and its first peak appeared at 93 ms in the right (contralateral) and at 125 ms in the left (ipsilateral) hemisphere. Such a latency difference could be due to callosal transmission time. Sporadically, there were some other higher-order sources active, some in the parietal and some in frontotemporal or deep frontal regions (which may indicate distributed superficial activity). Our data indicate that after V1 the LOV5 region activates within about 10 ms. The V2 was not separable in group data, but few individuals for whom this source was separate showed sequential activation in V2, with latencies between the V1 and V3 (Figure 3). The LOV5 ipsilateral to the stimulus activated some 40 ms later than V1. Earlier MEG studies have indicated very early parieto-occipital or precuneus activation both for pattern onset (Tzelepi et al., 2001) and for pattern reversal (Vanni et al., 2001b) stimuli. Such activation may be included in our V1+ and V3V3a+ sources' first deflection, because after lower visual field stimulation, activations in both V1 and V3 lie dorsally, essentially in front of the parieto-occipital region, causing difficulties for separation of a source behind them. This parieto-occipital region also generates most alpha rhythm in humans (Williamson & Kaufman, 1989; Hari & Salmelin, 1997; Vanni, Revonsuo, & Hari, 1997), and perhaps changes in synchrony here are reflected in EEG and MEG, but not in fMRI, which would leave us with no a priori seed. In addition to the sources mentioned above, both our EEG recordings, earlier MEG recordings, and our control fMRI data in some subjects show additional activation in higher-order areas in parietal, temporal and frontal lobes, mainly contralateral to the stimulated visual field. It seems as if there would be a network of visual information processing areas, which become active always, despite our passive paradigm and rather meaningless checkerboard stimuli.

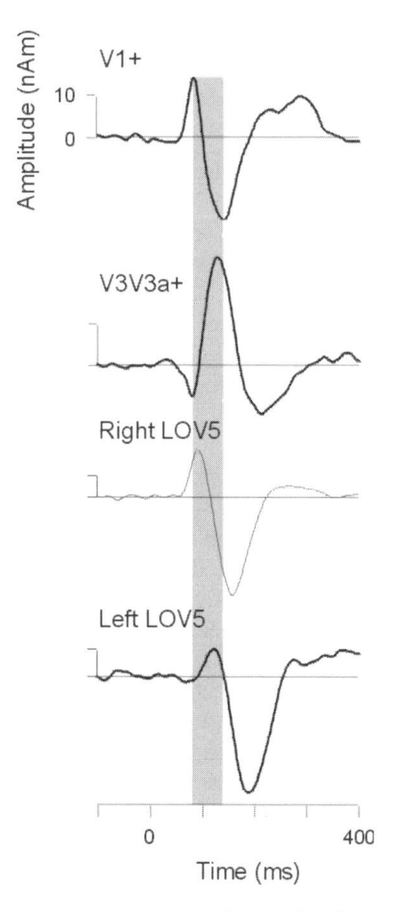

Figure 4: Group average source waveforms for the areas which could be separated in most subjects. Some activation at higher order areas has been omitted. The gray area indicates time from 85 to 140 ms.

Conclusions

We know from monkey data that the timing of information processing is not as assumed from hierarchical models. In addition to the unequal timing across anatomical hierarchy, activation latencies are not always fixed. The stimulus and task parameters may affect the dynamics in the cortical network. Since the human visual system is considerably more spread out over larger areas than in monkeys, thus increasing the distances between functional areas, it is of interest to determine the succession of activation in humans. We applied position and orientation information from fMRI to EEG multidipole models, and were able to separate activation at the retinotopic areas V1 to V3a from areas higher in the anatomical hierarchy. Our results show that activation spreads very early over the retinotopic areas, and largest deflections at the retinotopic areas appear after robust activation at the lateral occipital or occipito-temporal regions. When applied with care, multimodal imaging can open human retinotopic and other visual areas for quantitative studies of temporal dynamics.

References

Amunts, K., Malikovic, A., Mohlberg, H., Schormann, T., & Zilles, K. (2000). Brodmann's areas 17 and 18 brought into stereotaxic space—where and how variable. *Neuroimage, 11*, 66–84.

Boynton, G. M., Engel, S. A., Glover, G. H., & Heeger, D. J. (1996). Linear systems analysis of functional magnetic resonance imaging in human V1. *Journal of Neuroscience, 16*, 4207–4221.

Bullier, J. (2001). Integrated model of visual processing. *Brain Research/ Brain Research Reviews, 36*, 96–107.

Chen, W., & Ogawa, S. (1999). Principles of BOLD functional imaging. In C. T. W. Moonen & P. A. Bandettini (Eds.), *Functional MRI* (pp. 103–113). Berlin: Springer.

Dale, A. M., Liu, A. K., Fischl, B. R., Buckner, R. L., Belliveau, J. W., Lewine, J. D., & Halgren, E. (2000). Dynamic statistical parametric mapping: combining fMRI and MEG for high-resolution imaging of cortical activity. *Neuron, 26*, 55–67.

Di Russo, F., Martinez, A., Sereno, M. I., Pitzalis, S., & Hillyard, S. A. (2002). Cortical sources of the early components of the visual evoked potential. *Human Brain Mapping, 15*, 95–111.

Disbrow, E. A., Slutsky, D. A., Roberts, T. P., & Krubitzer, L. A. (2000). Functional MRI at 1.5 tesla: A comparison of the blood oxygenation level-dependent signal and electrophysiology. *Proceedings of the National Academy of Science of the USA, 97 (17)*, 9718–9723.

Foxe, J. J., & G. V. Simpson (2002). Flow of activation from V1 to frontal cortex in humans. A framework for defining "early" visual processing. *Experimental Brain Research, 142*, 139–150.

Girard, P., Hupe, J. M., & Bullier, J. (2001). Feedforward and feedback connections between areas V1 and V2 of the monkey have similar rapid conduction velocities. *Journal of Neurophysiology, 85*, 1328–11331.

Hari, R. (1990). The neuromagnetic method in the study of the human auditory cortex. In F. Grandori, M. Hoke, & G. Romani (Eds.), *Auditory evoked magnetic fields and potentials. Advances in audiology*. Volume 6 (pp. 222–282). Basel: Karger.

Hari, R., & Salmelin, R. (1997). Human cortical oscillations: A neuromagnetic view through the skull. *Trends in Neurosciences, 20*, 44–49.

Hoge, R. D., Atkinson, J., Gill, B., Crelier, G. R., Marrett, S., & Pike, G. B. (1999). Linear coupling between cerebral blood flow and oxygen consumption in activated human cortex. *Proceedings of the National Academy of Science of the USA*, 96 (16), 9403–9408.

Hupe, J. M., James, A. C., Girard, P., Lomber, S. G., Payne, B. R., & Bullier, J. (2001). Feedback connections act on the early part of the responses in monkey visual cortex. *Journal of Neurophysiology, 85*, 134–145.

Hupe, J. M., James, A. C., Payne, B. R., Lomber, S. G., Girard, P., & Bullier, J. (1998). Cortical feedback improves discrimination between figure and background by V1, V2 and V3 neurons. *Nature, 394*, 784–787.

Hämäläinen, M., Hari, R., Ilmoniemi, R., Knuutila, J., & Lounasmaa, O. V. (1993). Magnetoencephalography: Theory, instrumentation, and applications to noninvasive studies of the working human brain. *Reviews of Modern Physics, 65*, 413–497.

James, A. (2003). A pattern pulse multifocal visual evoked potential. *Investigative Ophtalmology and Vision Science*, 44, 879–890.

Kaneoke, Y., Bundou, M., & Kakigi, R. (1998). Timing of motion representation in the human visual system. *Brain Research, 790*, 195–201.

Liu, A. K., Belliveau, J. W., & Dale, A. M. (1998). Spatiotemporal imaging of human brain activity using functional MRI constrained magnetoencephalography data: Monte Carlo simulations. *Proceedings of the National Academy of Science of the USA, 95 (15)*, 8945–8950.

Logothetis, N. K., Pauls, J., Augath, M., Trinath, T., & Oeltermann, A. (2001). Neurophysiological investigation of the basis of the fMRI signal. *Nature, 412*, 150–157.

Lütkenhöner, B. (1998). Dipole separability in a neuromagnetic source analysis. *IEEE Transactions on Biomedical Engineering, 45*, 572–581.

Lütkenhöner, B., Greenblatt, R., Hämäläinen, M., Mosher, J., Scherg, M., Tesche, C., & Valdes Sosa, P. (2000). Comparison between different approaches to the biomagnetic inverse problem – workshop report. In C. J. Aine, Y. Okada, G. Stroink, S. Swithenby, & C. Wood (Eds.), *Proceedings of the tenth international conference on biomagnetism* (pp. 163–176). New York: Springer.

McGonigle, D. J., Howseman, A. M., Athwal, B. S., Friston, K. J., Frackowiak, R. S., & Holmes, A. P. (2000). Variability in fMRI: an examination of intersession differences. *Neuroimage, 11* (6 Pt. 1), 708–734.

Mosher, J. C., M. E. Spencer, R. M. Leahy, & P. S. Lewis (1993). Error bounds for EEG and MEG dipole source localization. *Electroencephalography and Clinical Neurophysiology, 86*, 303–321.

Nowak, L. G., & Bullier, J. (1997). The timing of information transfer in the visual system. In K. S. Rockland, J. H. Kaas, & A. Peters (Eds.), *Cerebral Cortex*. Volume 12 (pp. 205–241). New York: Plenum Press.

Portin, K., Vanni, S., Virsu, V., & Hari, R. (1999). Stronger occipital cortical activation to lower than upper visual field stimuli. Neuromagnetic recordings. *Experimental Brain Research, 124*, 287–294.

Price, C. J., & Friston, K. J. (1997). Cognitive conjunction: a new approach to brain activation experiments. *Neuroimage*, 5 (4 Pt. 1), 261–270.

Salmelin, R., Hari, R., Lounasmaa, O., & Sams, M. (1994). Dynamics of brain activation during picture naming. *Nature, 368*, 463–465.

Scherg, M., & Berg, P. (1991). Use of prior knowledge in brain electromagnetic source analysis. *Brain Topography, 4*, 143–150.

Schmolesky, M. T., Wang, Y., Hanes, D., Thompson, K. G., Leutgeb, S., Schall, J. D., & Leventhal, A. G. (1998). Signal timing across macaque visual system. *Journal of Neurophysiology, 79*, 3272–3278.

Schroeder, C. E., Mehta, A. D., & Givre, S. J. (1998). A spatiotemporal profile of visual system activation revealed by current source density analysis in the awake macaque. *Cerebral Cortex, 8*, 575–592.

Srinivasan, R., Tucker, D. M., & Murias, M. (1998). Estimating the spatial Nyquist of the human EEG. *Behavior Research Method, Instruments, and Computers, 30*, 8–19.

Stenbacka, L., Vanni, S., Uutela, K., & Hari, R. (2002). Comparison of minimum current estimate and dipole modeling in the analysis of simulated activity in the human visual cortices. *Neuroimage*, *16*, 936.

Tesche, C. D., Uusitalo, M. A., Ilmoniemi, R. J., Huotilainen, M., Kajola, M., & Salonen, O. (1995). Signal-space projections of MEG data characterize bothdistributed and well-localized neuronal sources. *Electroencephalography and Clinical Neurophysiology*, *95*, 189–200.

Thorpe, S., Fize, D., & Marlot, C. (1996). Speed of processing in the human visual system. *Nature*, *381*, 520–522.

Tzelepi, A., Ioannides, A. A., & Poghosyan, V. (2001). Early (N70m) neuromagnetic signal topography and striate and extrastriate generators following pattern onset quadrant stimulation. *Neuroimage*, *13*, 702–718.

Uusitalo, M. A., & Ilmoniemi, R. J. (1997). Signal-space projection (SSP) method for separating MEG or EEG into components. *Medical and Biological Engineering and Computing*, *35*, 135–140.

Uutela, K., Hämäläinen, M., & Somersalo, E. (1999). Visualization of magnetoencephalographic data using minimum current estimates. *Neuroimage*, *10*, 173–180.

Vanni, S., Dojat, M., Warnking, J., Delon-Martin, C., Segebarth, C., & Bullier, J. (submitted-a). Timing of intaractions across the visual field in human cortex.

Vanni, S., Raninen, A., Nasanen, R., Tanskanen, T., & Hyvarinen, L. (2001a). Dynamics of cortical activation in a hemianopic patient. *Neuroreport*, *12*, 861–865.

Vanni, S., Revonsuo, A., & Hari, R. (1997). Modulation of the parieto-occipital alpha rhythm during object detection. *Journal of Neuroscience*, 17, 7141–7147.

Vanni, S., Revonsuo, A., Saarinen, J., & Hari, R. (1996). Visual awareness of objects correlates with activity of right occipital cortex. *Neuroreport*, *8*, 183–186.

Vanni, S., Tanskanen, T., Seppa, M., Uutela, K., & Hari, R. (2001b). Coinciding early activation of the human primary visual cortex and anteromedial cuneus. *Proceedings of the National Academy of Science of the USA*, 98 (5), 2776–2780.

Vanni, S., & Uutela, K. (2000). Foveal attention modulates responses to peripheral stimuli. *Journal of Neurophysiology*, *83*, 2443–2452.

Vanni, S., Warnking, J., Dojat, M., Delon-Martin, C., Bullier, J., & Segebarth, C. (submitted-b). Sequence of pattern onset responses in the human visual areas.

von Helmholtz, H. (1853). Über einige Gesetze der Vertheilung elektrischer Ströme in körperlichen Leitern, mit Anwendung auf die thierisch-elektrischen Versuche. *Annals of Physical Chemistry*, *89*, 353–377.

Williamson, S., & Kaufman, L. (1989). Advances in neuromagnetic instrumentation and studies of spontaneous brain activity. *Brain Topography*, *2*, 129–139.

Chapter 9:
Time and conscious visual processing

ANDREAS K. ENGEL

Abstract

Cognitive functions like perception, memory, language or consciousness are based on highly parallel and distributed information processing by the brain. One of the major unresolved questions is how information can be integrated and how coherent representational states can be established in the distributed neuronal systems subserving these functions. It has been suggested that this so-called "binding problem" may be solved in the temporal domain. The hypothesis is that synchronization of neuronal discharges can serve for the integration of distributed neurons into cell assemblies and that this process may underlie the selection of perceptually and behaviorally relevant information. The chapter will review experimental results, obtained in sensory systems of both animals and humans, which support this notion of temporal binding and suggest that synchrony is related to the buildup of states of sensory awareness.

Introduction: Binding and awareness

This chapter intends to contribute to the ongoing debate about the neural correlates of consciousness from the viewpoint of a particular experimental approach: the study of distributed neuronal processing and of dynamic interactions which implement specific bindings in neural network architectures. The notion of binding and the search for potential binding mechanisms has received increasing attention during the past decade. Having been introduced first in the psychological discourse, the issue of binding has now advanced into the focus of research also in other disciplines within cognitive science such as neural network modeling, philosophy of mind and cognitive neuroscience.

In all these domains, the problem has been identified that encoding and retrieval of information in neuronal networks requires some sort of mechanism which allows the expression of specific relationships between elementary processors. This so-called binding problem arises for several reasons: First, information processing underlying cognitive functions is typically distributed across many network elements and, thus, one needs to identify those neurons or network nodes that currently participate in the same cognitive process. Second, perception of and action in a complex environment usually requires the parallel processing of information related to different objects or events that have to be kept apart to allow sensory segmentation and goal-directed behavior. Thus, neuronal activity pertaining e.g. to a particular object needs to be distinguished from unrelated information in order to avoid confusion and erroneous conjunctions (von der Malsburg, 1981). Third, it has been claimed that specific yet flexible binding is required within distributed activation patterns to allow the generation of syntactic structures and to account for the systematicity and productivity of cognitive processes (Fodor & Pylyshyn, 1988). Fourth, many cognitive functions imply the context-dependent selection of relevant information from a richer set of available data. It has been suggested that appropriate binding may be a prerequisite for the selection and further joint processing of subsets of information (Singer et al., 1997; Singer, 1999). These arguments suggest that cognitive functions require the implementation of binding mechanisms in the distributed networks subserving these functions.

In what follows, I want to focus on the idea that some kind of binding mechanism may also be critical for the establishment of conscious mental states. In recent years, several authors have emphasized a close link between binding and consciousness, following the intuition that consciousness requires some kind of integration, or coherence, of mental contents. Damasio (1990) has suggested that conscious recall of memory contents requires the binding of distributed information stored in spatially separate cortical areas. In various publications, Crick and Koch have discussed the idea that binding may be intimately related to the neural mechanisms of sensory awareness (Crick & Koch, 1990). According to their view, only appropriately bound neuronal activity can enter short-term memory and, hence, become available for access to phenomenal consciousness. Llinas and coworkers (1994) have proposed that arousal and awareness require binding of sensory information which is implemented by interactions between specific and nonspecific thalamocortical loops. Metzinger (1995) has extended this discussion by speculating that binding mechanisms might not only account for low-level properties of phenomenal consciousness like the holistic character of perceptual objects, but also for the formation of a phenomenal self-model and its embedding into a global world-model.

In the present paper, the discussion will be restricted to one particular aspect of consciousness, namely, sensory awareness. With many authors, I share the view that sensory awareness is one of those facets of consciousness that is probably most easily accessible both in terms of experimental quantification and theoretical explanation (Crick & Koch, 1990). Furthermore, there can be little doubt that we have this basic form of phenomenal consciousness in common with many other species (presumably with at least most other higher mammals). Thus, it is conceivable that research on animals can contribute substantially to explaining this aspect of consciousness, which may not hold for many higher-order features of consciousness which, for instance, require a language system or an elaborated self-model.

There seems to be wide agreement that awareness as the basic form of phenomenal consciousness has the following prerequisites (Crick & Koch, 1990): First, generating sensory awareness seems to involve some form of attentional mechanism, i.e., a mechanism that selects relevant information and enhances its impact on subsequent processing stages. Second, awareness presumably requires working memory, which allows the short-term storage of episodic contents. Third, awareness seems to presuppose the capacity for structured representation, i.e., the ability to achieve coherence of the contents of mental states and to establish specific relationships between representational items. The basic assumption made in the following is that all three capacities are, on the one hand, closely related to each other and, on the other hand, strongly dependent on binding mechanisms implemented in sensory systems. As one particular candidate for the latter, I will discuss dynamic binding by transient and precise synchronization of neuronal discharges (Engel, Roelfsema, Fries, Brecht, & Singer, 1997; Engel, Fries, & Singer, 2001; Engel & Singer 2001; Singer et al., 1997; Singer, 1999). As I will argue, there is now empirical evidence suggesting that temporal binding may be crucial for generating functionally efficacious representational states and for the selection of perceptually or behaviorally relevant information.

The concept of temporal binding

The concept of dynamic binding by synchronization of neuronal discharges has been developed mainly in the context of perceptual processing. One source of inspiration for this model has come from the insight that perception, like most other cognitive functions, is based on highly parallel information processing carried out by numerous brain areas. A paradigmatic case is provided by visual processing which shows a highly distributed organization. In monkeys, anatomical and physiological studies have led to the identification of more than 30 distinct visual areas in the cortex (Felleman & Van Essen, 1991). This parcellation is assumed to reflect some kind of functional specialization since neurons in each of these visual areas are, at least to some degree, selective for characteristic subsets of object features. Thus, for instance, some areas contain cells responding to the color of objects, while others primarily process information about the form of an object or its direction of motion in the visual field. As a consequence of this functional specialization, any object present in the field of view will activate feature-detecting neurons in many cortical areas simultaneously. The highly complex organization of visual processing naturally raises the question of how distributed neuronal responses can be integrated, which seems necessary to enable the brain to represent and store information about the external world in a useful way.

It has been suggested that the binding problem arising in distributed sensory networks may be solved by a mechanism which exploits the temporal aspects of neuronal activity (von der Malsburg, 1981; for review, see Engel et al., 1997; Singer et al., 1997; Singer, 1999). The prediction is that neurons which respond to the same sensory object might fire their action potentials in temporal synchrony (with a precision in the millisecond range). However, no such synchronization should occur between cells which are activated by different objects appearing in sensory space. Such a temporal integration mechanism would provide an elegant solution to the binding problem since, on the one hand, the synchrony would selectively tag the responses of neurons that code for the same object and demarcate their responses from those of neurons activated by other objects. This highly selective temporal structure would allow to establish a distinct representational pattern (a so-called assembly) for each object and, thus, would enable the visual system to achieve figure-ground segregation. On the other hand, such a temporal binding mechanism could also serve to establish relationships between neuronal responses over large distances and, thus, solve the problems imposed by the anatomical segregation of specialized processing areas.

Figure 1 schematically illustrates this temporal binding model. The example chosen is a bistable figure (Figure 1a). "Bistability" means that two interpretations are possible–here either the percept of one face that is partially occluded by a candlestick (Figure 1b) or, alternatively, two opposing faces (Figure 1c). These perceptual situations mutually exclude each other, and most observers flip back and forth between the two. For this example, the temporal binding model predicts that neurons should dynamically switch between assemblies and, hence, temporal correlations should differ for the two perceptual states. To illustrate this point, four visual cortical neurons are considered with receptive fields (circles 1–4 in Figure 1a) over image components whose grouping changes with the transition from one percept to the other. As shown schematically in Figure 1d, neurons 1 and 2 should synchronize if the respective contours are part of the one background face, and the same should hold for neurons 3 and 4 which represent contours of the candlestick. However, when the image is segmented into two opposing faces, the temporal coalition switches to synchrony between 1–3 and 2–4, respectively (Figure 1e). The assumption is that synchrony among neurons is subject to both bottom-up and top-down influences. On the one hand, binding within the respective objects is facilitated by Gestalt criteria that lead to contour grouping. On the other hand, factors like expectation, attention or previous knowledge about the objects encountered is likely to be crucial for the segmentation process. In the example chosen here, switching between the two perceptual constellations in terms of synchrony cannot be explained by reference to low-level Gestalt criteria, but presumably requires interactions with higher-level assemblies carrying invariant object representations. Very likely, competition among such populations representing high-level contents must also be involved.

This strategy of temporal binding exhibits a number of crucial advantages. First, it preserves the general advantages of distributed coding schemes such as robustness against loss of network elements and richness of representations which contain explicit information about object features and do not only signal the presence of the object. Second, this strategy enhances processing speed because binding can, in principle, occur using the very first spikes of a response (Singer et al., 1997). Third, temporal binding alleviates superposition problems that occur in conventional distributed systems that operate solely on the basis of average firing rates (von der Malsburg, 1981). The reason is that using synchrony as an additional coding dimension allows the dissociation of the binding code from the feature code (object features being signaled by firing rates). This allows to coactivate multiple assemblies without confusion, because the temporal relationship between neuronal discharges permits the unambiguous distinction of subsets of functionally related responses (see Figure 1). Fourth, temporal binding provides an efficient mechanism for selection of assemblies for further processing (Singer et al., 1997; Singer, 1999), because precisely synchronized spikes constitute highly salient events which can be detected by coincidence-sensitive neurons in other brain areas (König, Engel, & Singer, 1996).

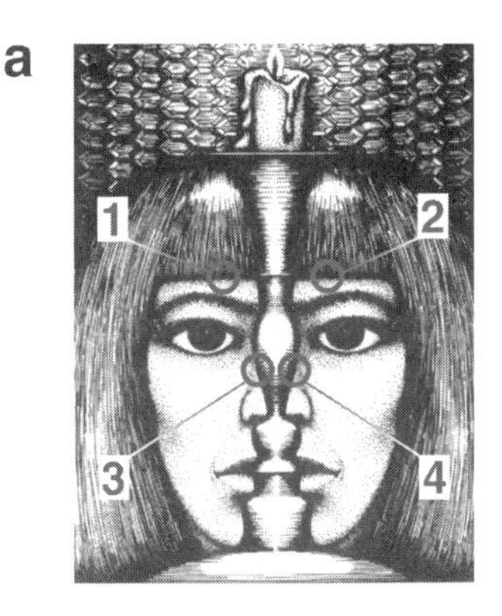

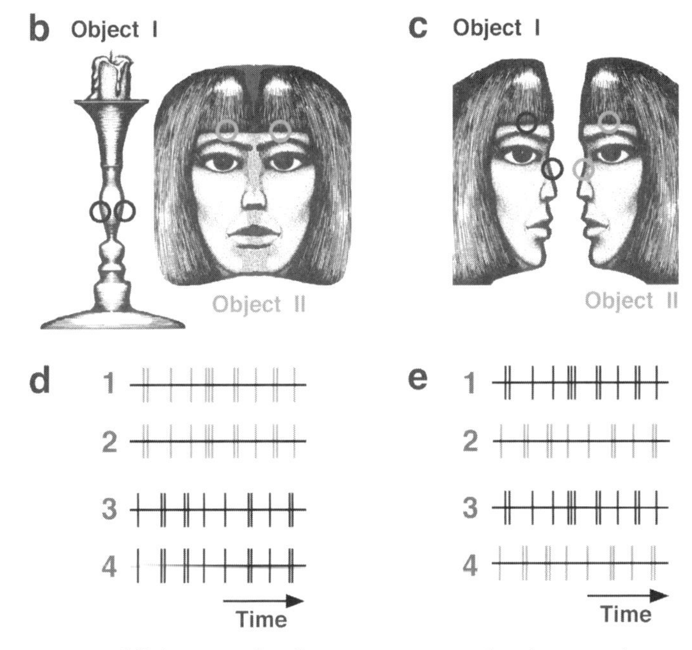

Figure 1: Establishment of coherent representational states by temporal binding. (a) Bistable figure and receptive fields (1–4) of four hypothetical visual cortical neurons. (b) Percept of one face that is partially occluded by a candlestick. (c) Percept of two opposing faces. (d) Neurons 1 and 2 should synchronize if the respective contours are part of the one background face, and the same should hold for neurons 3 and 4 which represent contours of the candlestick. (e) When the image is segmented into two opposing faces, the temporal coalition is assumed to switch to synchrony between 1–3 and 2–4, respectively. See text for further explanations. Panel a is adapted with modification from Shepard (1990)

In the present context, the most spectacular extension of the concept of temporal binding has been its application to the problem of consciousness by Crick and Koch. As has been mentioned already, they have argued for a close relationship between binding and sensory awareness (Crick & Koch, 1990). Beyond that, they were the first to suggest that it could be a temporal binding mechanism of the kind discussed here which is required for the establishment of awareness. Inspired by the finding that visual stimuli can elicit synchronized oscillatory activity in the visual cortex (Eckhorn et al., 1988; Gray & Singer, 1989; Gray, König, Engel, & Singer, 1989), they proposed that an attentional mechanism induces synchronous oscillations in selected neuronal populations, and that this temporal structure would facilitate transfer of the encoded information to working memory. The provocative scent of this hypothesis comes from the authors' implicit assumption that these are not just necessary, but indeed sufficient conditions for the occurrence of awareness. At the time it was published, Crick and Koch's speculative proposal was not supported by experimental evidence. In later sections of this chapter, I will discuss more recent data which suggest that temporal binding may indeed be a prerequisite for the access of information to phenomenal consciousness. However, although largely in line with Crick and Koch's hypothesis, the present data do not seem to support the conclusion that synchronization of assemblies would constitute a *sufficient* condition for production of awareness.

Physiological evidence for temporal binding

By now, the synchronization phenomena predicted by the temporal binding hypothesis are well documented for a wide variety of neural systems. It is well established that neurons in both cortical and subcortical centers can synchronize their discharges with a precision in the millisecond range (for review, see Engel et al., 1997; Engel et al., 2001; Singer et al., 1997; Singer, 1999). This has been demonstrated in particular for the visual system, but similar observations have been made for the other sensory systems, for the motor system, and for cortical association areas. In the following, we will focus on experimental data suggesting that the observed synchrony does indeed serve for the binding and selection of functionally related responses. These data have been obtained mainly in experiments on cats and monkeys, but presumably the results can be generalized to the human brain where recent EEG and MEG studies have provided evidence for similar synchronization phenomena.

For the case of the visual system, the temporal binding model predicts a synchronization of spatially separate cells within individual visual areas to account for the integration of perceptual information across different locations in the visual field. In addition, synchrony should occur across large distances in the cortex to allow for binding between visual areas involved in the analysis of different object features. According to the temporal binding model, this would be required for the full representation of objects. Both predictions have been confirmed experimentally. In cats and monkeys, synchrony has been observed within striate and extrastriate visual areas (e.g. Eckhorn et al., 1988; Gray & Singer, 1989; Kreiter & Singer, 1996). Moreover, it has been shown that response synchronization can extend well beyond the borders of a single visual area. Thus, for instance, correlated firing has been observed between neurons located in different cerebral hemispheres (Engel, König, Kreiter, & Singer, 1991a). In terms of the temporal binding hypothesis, this result is important because interhemispheric synchrony is required to bind the features of objects extending across the midline of the visual field. In addition, temporal correlations have been studied for neurons located in different areas of the same hemisphere. Finally, recent evidence shows that synchronous firing is not confined to the cortex but occurs also in subcortical visual structures such as the retina, the lateral geniculate nucleus and the superior colliculus (Neuenschwander & Singer, 1996; Brecht, Singer, & Engel, 1998, 1999).

Studies in non-visual sensory modalities and in the motor system have provided evidence for very similar synchronization phenomena. Synchrony is well known to occur in the olfactory system of various vertebrate and invertebrate species, where these phenomena have been related to the processing of odor information (Laurent, 1996). Moreover, in both the auditory (deCharms & Merzenich, 1996) and the somatosensory cortex (Murthy & Fetz, 1992) precise neuronal synchronization has been observed. Furthermore, neuronal interactions with a precision in the millisecond range have been described in the hippocampus (Buzsáki & Chrobak, 1995) and in the frontal cortex (Vaadia et al., 1995). Finally, similar evidence is available for the motor system where neural synchronization has been discovered both during preparation and execution of movements (Murthy & Fetz, 1992).

Although the temporal binding model offers an attractive conceptual scheme for understanding the binding and selection of distributed neuronal responses, definitive evidence that the brain actually uses synchronization in exactly this way has not yet been obtained. However, a number of findings strongly suggest that the synchrony is indeed functionally relevant. One important result supporting the temporal binding model is that neuronal synchronization in the visual system depends on the stimulus configuration (see Figure 2). It could be demonstrated that spatially separate cells show strong synchronization only if they respond to the same visual object. However, if responding to two independent stimuli, the cells fire in a less correlated manner or even without any fixed temporal relationship (Gray et al. 1989; Engel, Kreiter, König, & Singer, 1991b; Kreiter & Singer 1996). The experiments demonstrate that Gestalt criteria such as continuity or coherent motion, which have psychophysically been shown to support perceptual grouping, are important for the establishment of synchrony among neurons in the visual cortex. These data strongly support the hypothesis that correlated firing provides a dynamic mechanism which permits binding and response selection in a flexible manner.

Additional evidence that neuronal synchronization is indeed functionally relevant and related to the animal's perception is provided by experiments on cats with convergent squint (Roelfsema, König, Engel, Sireteanu, & Singer, 1994). Subjects with this type of strabismus often use only one eye for active fixation. The non-fixating eye then develops a perceptual deficit called strabismic amblyopia.

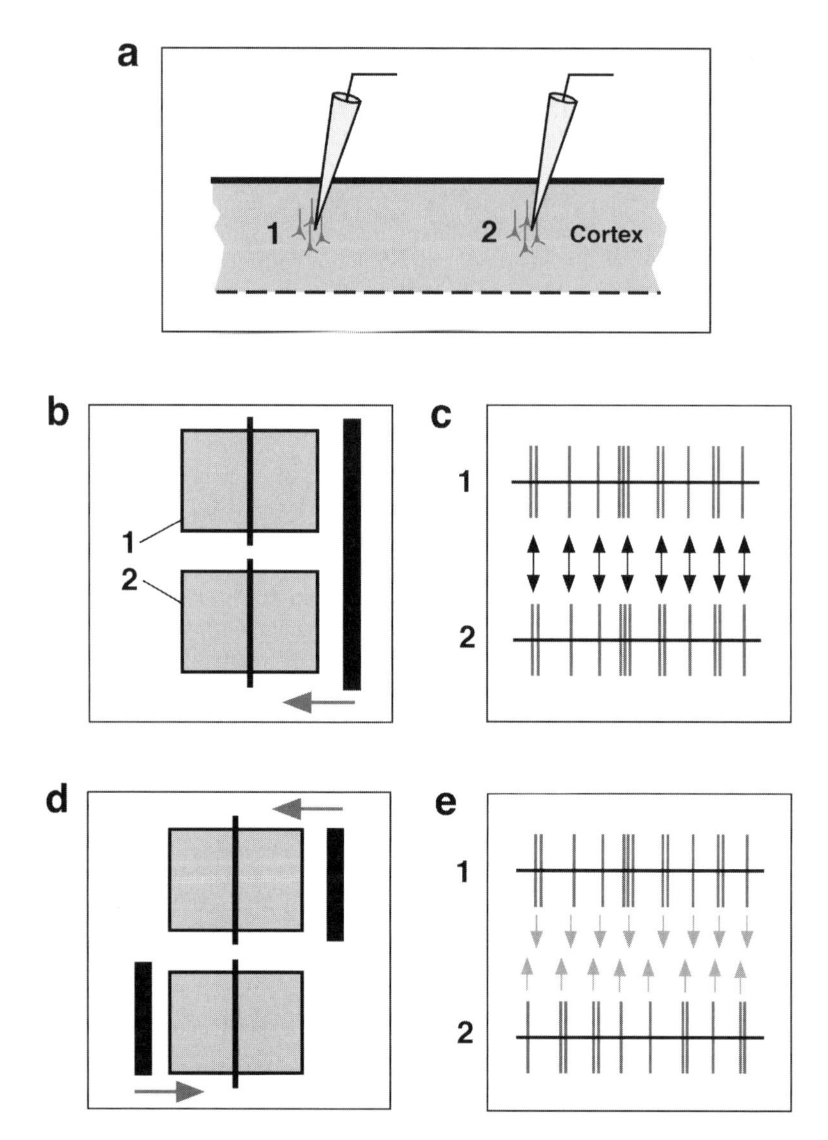

Figure 2: Synchrony in visual cortex is dependent on the configuration of visual stimuli. (a) Typically, activity of spatially separate cell groups (1, 2) is recorded with two microelectrodes. The neurons can then be activated using different stimulus paradigms. (b, c) If a single coherent object is moved across the receptive fields (1, 2), the cells at the two recording sites are synchronously active (arrows in c). (d, e) Activation of the same neurons with two different objects moving in opposite directions does not induce synchrony (cf. offset of arrowheads in e).

Symptoms of strabismic amblyopia include a reduced acuity of the affected eye, temporal instability and spatial distortions of the visual image, and the so-called crowding phenomenon, i.e., discrimination of details deteriorates further if other contours are nearby. Clearly, at least some of these deficits indicate a reduced capacity of integrating visual information and an impairment of the mechanisms responsible for feature binding. The results of the correlation study by Roelfsema et al. (1994) indicate that these perceptual deficits may be due to a disturbance of intracortical interactions. Thus, clear differences were observed in the synchronization of cells driven by the normal and the amblyopic eye, respectively (see Figure 3). In the primary visual cortex, responses of neurons activated through the amblyopic eye showed a much weaker correlation than the discharges of neurons driven by the normal eye. Surprisingly, however, in terms of average firing rates the responses of neurons driven by the normal and amblyopic eye were indistinguishable. These results indicate that strabismic amblyopia is accompanied by a selective impairment of intracortical interactions that synchronize neurons responding to coherent stimuli. As mentioned above, most of the problems in amblyopic vision result from an improper segregation of features and the formation of false conjunctions. Therefore, the fact that the only measurable abnormality correlating with the perceptual deficit was the reduced synchronicity is in good agreement with the hypothesis that synchronization is employed for feature binding and serves to disambiguate distributed response patterns.

Strong evidence for a relation to perceptual processing is also provided by recent experiments in awake behaving macaque monkeys. These studies demonstrate that synchrony among cortical neurons is enhanced during attentional selection of sensory information. Steinmetz et al. (2000) have investigated cross-modal attentional shifts in awake monkeys that had to direct attention to either visual or tactile stimuli that were presented simultaneously. Neuronal activity was recorded in the secondary somatosensory cortex. For a significant fraction of the neuronal pairs in this area, synchrony depended strongly on the monkey's attention. If the monkey shifted attention to the visual task, temporal correlations typically decreased among somatosensory cells, as compared to task epochs where attention was not distracted from the somatosensory stimuli. Along a similar vein, strong attentional effects on temporal response patterning have also been observed in monkey V4 (Fries, Reynolds, Rorie, & Desimone, 2001). In this study, two stimuli were presented simultaneously on a screen, one inside the receptive fields of the recorded neurons and the other nearby. The animals had to detect subtle changes in one or the other stimulus. If attention was shifted towards the stimulus processed by the recorded cells, there was a marked increase in local synchronization.

Evidence for a functional role of neural synchrony is also provided by recent studies of sensorimotor interactions. Synchronization between sensory and motor assemblies has been investigated in a recent study on awake behaving cats that were trained to perform a visuomotor coordination task (Roelfsema, Engel, König, & Singer, 1997). In these animals, neural activity was recorded with electrodes chronically implanted in various areas of the visual, parietal and motor cortex.

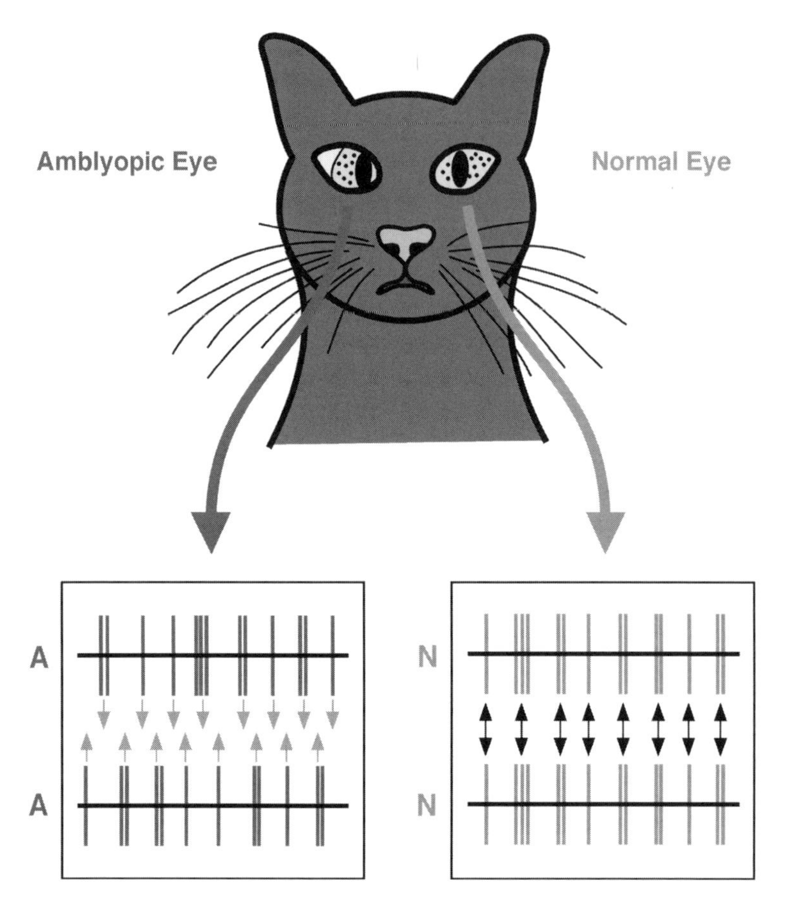

Figure 3: Neuronal synchronization in the primary visual cortex of animals with strabismic ambylopia. Experiments were performed on cats with a convergent squint (in this case an inward deviation of the animals right eye). The deviating eye develops a specific perceptual deficit denoted as strabismic amblyopia. The lower panel illustrates the temporal correlation between neurons driven by the amblyopic eye (A, left) and the normal eye (N, right), respectively. Temporal correlation is strong if both recording sites are driven by the normal eye (arrows). Synchronization is, however, much weaker or absent between cells dominated by the amblyopic eye (A).

The results of this study show that synchronization of neural responses does not only occur within the visual system, but also between visual and parietal areas as well as between parietal and motor cortex. Importantly, the interareal interactions changed dramatically in different behavioral situations. Precise neuronal synchronization between sensory and motor areas occurred specifically in those task epochs where the animal had to process visual information attentively to direct the required motor response. The observations of this study suggest that synchrony may indeed be relevant for visuomotor coordination and may serve for the linkage of sensory and motor aspects of behavior. The specificity of such interactions might allow, for instance, the selective channeling of sensory information to different motor programs which are concurrently executed. Similar conclusions are suggested by studies in monkeys, where synchronization between sensory and motor cortical areas has also been reported (Murthy & Fetz, 1992).

Another example for the potential relevance of synchrony for sensorimotor transformations is provided by interaction of cortical assemblies with subcortical integrative structures such as the superior colliculus. Recent experiments in the cat show that neurons in visual cortical areas can synchronize, via the corticotectal pathway, with cells in the superficial layers of the colliculus (Brecht et al., 1998, 1999). These studies have revealed the occurrence of precise temporal relationships between cortical and collicular neurons. Moreover, it could be shown that corticotectal interactions are strongly dependent on the temporal coherence of cortical activity. This finding is consistent with the idea that the temporal organization of activity patterns determines the efficiency of the output of the visual cortex. More recent experiments have directly tested the role of synchrony for the selection of targets during orienting responses of the animal, which are mediated by the superior colliculus (Brecht, Singer, & Engel, 1997). In these experiments, it was investigated how electrically evoked saccadic eye movements were affected by varying the temporal relation between microstimulation trains applied at two different sites in the colliculus. Synchronous activation of two collicular sites led to vector averaging, i.e., to movements along a vector corresponding to the mean of the saccades evoked by stimulating the two sites individually. In contrast, asynchronous stimulation (10ms or 5ms offset between the pulses of the two stimulation trains) led to vector summation, i.e., in this case, the saccades had the same direction as those evoked by synchronous pulse trains, but showed approximately double amplitude. These data show that vector averaging in the colliculus is restricted to synchronously active cells, whereas small temporal phase-shifts lead to a radically different motor strategy. This strongly suggests that synchrony in the millisecond range is an important determinant for the motor output in sensorimotor loops that read the temporally encoded information from sensory assemblies.

Specific changes of neural synchronization associated with attention and perceptual processing are also demonstrated by EEG and MEG studies in humans. In many of the animal studies cited above, precise synchrony among distributed neurons is associated with an oscillatory modulation of the responses at frequencies in the gamma range, i.e., between 30 and 80 Hz (for a review, see Engel et al., 1997; Singer & Gray, 1995). Synchronization in the same frequency band has been shown to be associated with cognitive processing in humans (Tallon-Baudry & Bertrand, 1999; Engel & Singer, 2001). The occurrence of evoked gamma-band responses was demonstrated first in the auditory system (Galambos, Makeig, & Talmachoff, 1981; Pantev et al., 1991). Subsequently, occurrence of gamma frequency components has been studied also for visual (Tallon-Baudry, Bertrand, Delpuech, & Pernier, 1997; Müller, Junghöfer, Elbert, & Rockstroh, 1997; Rodriguez et al., 1999; Herrmann, Mecklinger, & Pfeifer, 1999) and language processing (Pulvermüller, Lutzenberger, Preissl, & Birbaumer, 1995) and during the execution of motor tasks (Kristeva-Feige, Feige, Makeig, Ross, & Elbert, 1993). What the studies demonstrate is the occurrence of gamma activity under a wide variety of tasks and paradigms including processing of coherent stimuli (see Figure 4), perceptual discrimination, focussed attention, short-term memory, and sensorimotor integration. Typically, the observed amount of gamma is positively correlated with increased "processing load" and, thus, with the level of attention, as well as with the difficulty or integrative nature of the processing. Generally, the human data are in good agreement with animal studies suggesting a role of gamma synchronization in the binding and selection of distributed information.

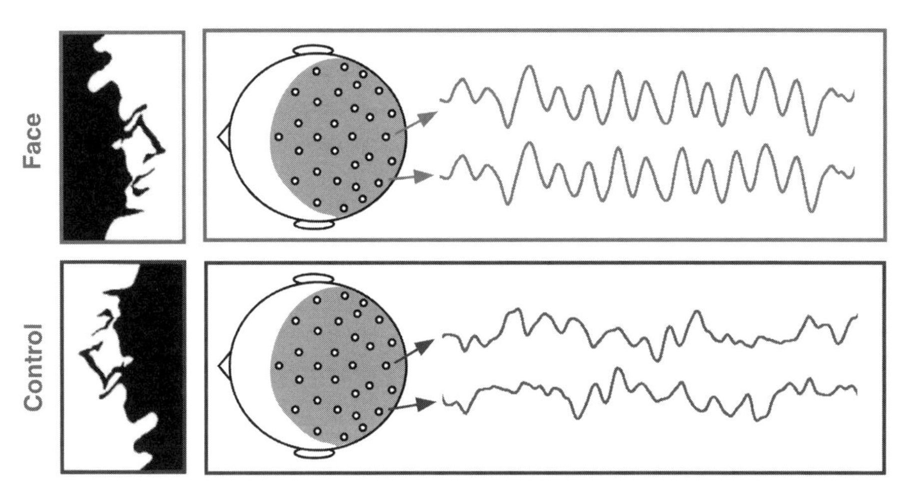

Figure 4: Neuronal synchrony as a correlate of perception in human subjects. In a recent study, Varela and coworkers studied the correlates of perceptual binding using EEG recordings (Rodriguez et al., 1999). White dots in the middle panels represent EEG-electrodes mounted to the head of the experimental subjects. During the measurements, images were shown that could be interpreted as faces when presented upright, but were not considered as meaningful when inverted (control). As shown schematically to the right, perception of faces was associated with fast EEG-waves that were phase-synchronized across different recording sites. Synchronized oscillations were particularly prominent during the epoch in time where conscious percepts were formed by the subjects (not shown). In the control condition (bottom), the same physical stimuli did not induce synchronized responses and fast oscillatory components were less prominent.

Binding and phenomenal consciousness

The experimental data discussed in the preceding section clearly argue for the importance of synchrony in the establishment of coherent sensory representations and for sensorimotor integration. Recent evidence indicates that these synchronization phenomena may also be relevant for the buildup of phenomenal states and the selection of visual information for access to awareness. This is suggested by experiments in which we recorded neuronal responses from the visual cortex of strabismic cats under conditions of binocular rivalry (Fries, Roelfsema, Engel, König, & Singer, 1997; Fries, Schröder, Roelfsema, Singer, & Engel, 2002). Binocular rivalry is a particularly interesting case of dynamic response selection which occurs when the images in the two eyes are incongruent and cannot be fused into a coherent percept. In this case, only signals from one of the two eyes are selected and perceived, whereas those from the other eye are suppressed. In normal subjects, perception alternates between the stimuli presented to left and right eye, respectively. The important point is that this shift in perceptual dominance can occur without any change of the physical stimulus. Obviously, this experimental situation is particularly revealing for the issue at stake, because neuronal responses to a given stimulus can be studied either with or without being accompanied by awareness (Crick & Koch, 1990) and, thus, there is a chance of revealing the mechanisms leading to the selection of perceptual information.

Previous studies have examined the hypothesis that response selection in binocular rivalry is achieved by a modulation of firing rate. In these experiments, a number of different visual cortical areas were recorded in awake monkeys experiencing binocular rivalry. With respect to early processing stages (the first, second, and fourth visual area and the middle temporal area MT), the results were not conclusive (Logothetis & Schall, 1989; Leopold & Logothetis, 1996). The fraction of neurons that decreased their firing rates upon suppression of the stimulus to which they responded was about the same as the fraction of cells that increased their discharge rate and altogether response amplitudes changed in less than 50% of the neurons when eye dominance switched. A clear and positive correlation between firing rate and perception was found only in inferotemporal cortex, i.e. at a relatively late stage of visual processing (Sheinberg & Logothetis, 1997).

In our study (Fries et al., 1997, 2002) we have investigated the hypothesis that response selection in early visual areas might be achieved by modulation of the synchronicity rather than the rate of discharges. These measurements were performed in awake cats with wire electrodes chronically implanted in areas 17 and 18. The animals were subjected to dichoptic visual stimulation, i.e., patterns moving in different directions were simultaneously presented to the left and the right eye, respectively (see Figure 5). Perceptual dominance for a given set of stimuli was inferred from the direction of eye movements induced by the drifting gratings (the so-called optokinetic nystagmus) which were recorded by periorbital electrodes (during correlation measurements, however, precautions were taken to minimize eye movements; for details, see Fries et al., 1997). As a baseline, neuronal responses were also recorded under monocular stimulation conditions. The results obtained with this experimental approach show that visual cortical neurons driven by the dominant and the suppressed eye, respectively, do neither differ in the strength nor in the synchronicity of their response to monocular visual stimulation. They show, however, striking differences with respect to their synchronization behavior when exposed to the rivalry condition (see Figure 5). Neurons representing the stimulus that wins in rivalry and is perceived increase their synchrony, whereas cells processing the suppressed visual pattern decrease their temporal correlation. However, no differences were noted under the rivalry condition for the discharge rates of cells responding to the dominant and the suppressed eye, respectively.

These results show that, in areas 17 and 18 of awake, strabismic cats, dynamic selection and suppression of sensory signals are associated with modifications of the synchrony rather than the rate of neuronal discharges. This suggests that at an early level of visual processing, it is the degree of synchronicity rather than the amplitude of responses that determines which of the input signals will be processed further and then support perception and oculomotor responses. Changes in synchronicity at early stages of processing are bound to result in changes of discharge rate at later stages. Thus, the rate changes observed with perceptual rivalry in higher cortical areas (Sheinberg & Logothetis, 1997) could be secondary to modifications of neuronal synchronization at lower levels of processing.

In the present context, the important conclusion from these experiments is that only strongly synchronized neuronal responses can contribute to awareness and conscious phenomenal states. The data suggest that activation of feature-detecting cells is, as such, not sufficient to grant access of the encoded information to consciousness (note, that the cells representing the suppressed stimulus are still well responding). Rather, to be functionally effective and to be selected for perception, neurons have to be strongly synchronized and bound into assemblies. In this respect, the data support the proposal by Crick and Koch (1990) that neuronal synchronization may be a necessary condition for the occurrence of awareness. Admittedly, this conclusion rests on the assumption that sensory awareness of the stimulus correlates well with the oculomotor behavior that we have used as an indirect measure of the cats perceptual state. However, this correspondence has been well established in humans, where a nearly perfect correlation has been found between the direction of the pursuit phase of the optokinetic nystagmus and the perceived direction of motion (Fox, Todd, & Bettinger, 1975), meaning that it is impossible, in the rivalry situation, to track one of the patterns with the eyes but consciously perceive the other.

The crucial advantage of temporal binding is that it could permit the rapid and reliable selection of perceptually or behaviorally relevant information. Because precisely synchronized discharges have a high impact on the respective postsynaptic cells, the information tagged by such a temporal label could be rapidly and preferentially relayed to other processing centers (Singer et al., 1997). As proposed here, such a process of response selection, which is based on temporal correlation among subsets of activated neurons, may be integral part of the mechanisms responsible for perceptual awareness (Engel & Singer, 2001; Engel et al. 2001).

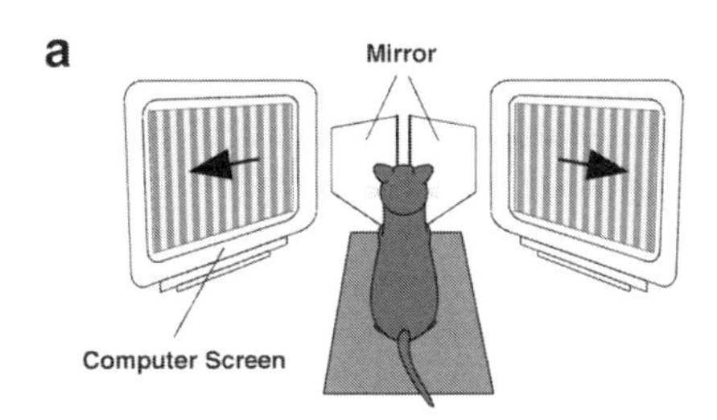

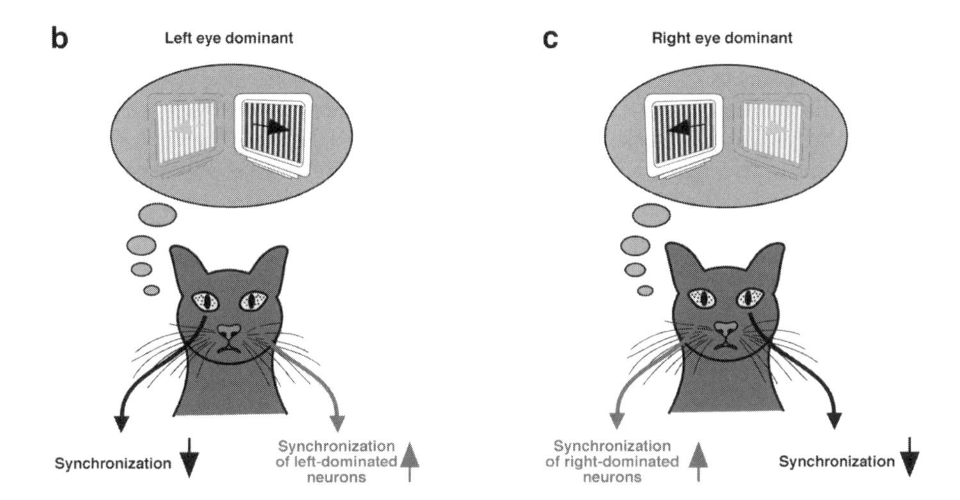

Figure 5: Neuronal synchronization under binocular rivalry. (a) To induce binocular rivalry, two mirrors are mounted in front of the animals head such that the eyes are viewing different stimuli. Under this condition, animals as well as humans alternate between two perceptual states. In certain episodes (b), the pattern presented to the left eye will dominate perception, while the information conveyed by the right eye is suppressed and excluded from perception. In other instances (c), the opposite effect may occur leading to perceptual dominance of the pattern presented to the right eye. As indicated in the bottom panel, synchrony will increase between neurons that represent the perceived stimulus, while it decreases between cells responding to the suppressed pattern. Thus, for instance, during dominance of the left eye (b), neurons driven by this eye will increase their temporal correlation, but correlation becomes weaker for cells driven by the right eye.

Conclusion

In this chapter, I have discussed the concept of temporal binding and its application to the issue of sensory processing and perceptual awareness. The basic assumption is that synchrony is introduced as an additional coding dimension which complements the conventional rate code. While the latter serves for the coarse coding of representational contents, the former may permit the dynamic expression of specific relations within a network. In this way, the combination of two different coding strategies could allow the multiplexing of different types of information within the same activity patterns and, thus, could enhance the representational power of distributed systems. As discussed above, the available data suggest that a temporal binding mechanism may indeed exist in the brain. Rather than being a futile epiphenomenon of network connectivity, precise synchronization of neuronal discharges seems to be functionally relevant for the binding of distributed responses in a wide variety of neural systems. Such a mechanism has the advantage that it raises jointly and with high temporal resolution the saliency of coherent subsets of signals, because downstream neuronal populations respond better to synchronized than to temporally dispersed spikes. For this reason, it is ideally suited to mediate selective transmission of coherent responses and, thus, for gating the access of relevant signals to phenomenal consciousness.

Taken together, these considerations suggest that studying the dynamics of neuronal interactions may be particularly rewarding in search for the neural correlates of consciousness. The important point of the results presented here is that, at least at early stages of sensory processing, the degree of synchronicity predicts reliably whether neural activity will contribute to conscious experience or not. Therefore, experiments designed to investigate neuronal synchronization may help to identify the selection mechanisms that are required for phenomenal consciousness. At this point, I have deliberately restricted discussion to the issue of awareness because it seems that, based on the present data, one can hardly argue about a relevance of binding mechanisms for other forms of consciousness. However, it has been speculated that temporal binding may also account for higher-order properties of phenomenal consciousness (Metzinger, 1995)—an exciting possibility that clearly awaits further research.

References

Brecht, M., Singer, W., & Engel, A. K. (1997). Collicular saccade vectors defined by synchronization. *Society for Neuroscience Abstracts, 23*, 843.

Brecht, M., Singer, W., & Engel, A. K. (1998). Correlation analysis of cortico-tectal interactions in the cat visual system. *Journal of Neurophysiology, 79,* 2394–2407.

Brecht, M., Singer, W., & Engel., A. K. (1999). Patterns of synchronization in the superior colliculus of anesthetized cats. *Journal of Neuroscience, 19*, 3567–3579.

Buzsáki, G., & Chrobak, J. J. (1995). Temporal structure in spatially organized neuronal ensembles: A role for interneuronal networks. *Current Opinion in Neurobiology, 5,* 504–510.

Crick, F., & Koch, C. (1990). Towards a neurobiological theory of consciousness. *Seminars in Neurosciences, 2*, 263–275.

Damasio, A. R. (1990). Synchronous activation in multiple cortical regions: a mechanism for recall. *Seminars in Neurosciences, 2*, 287–296.

deCharms, R. C., & Merzenich, M. M. (1996). Primary cortical representation of sounds by the coordination of action-potential timing. *Nature, 381*, 610–613.

Eckhorn, R., Bauer, R., Jordan, W., Brosch, M., Kruse, W., Munk, M., & Reitboeck, H. J. (1988). Coherent oscillations: A mechanism for feature linking in the visual cortex? *Biological Cybernetics, 60*, 121–130.

Engel, A. K., Fries, P., & Singer, W. (2001). Dynamic predictions: oscillations and synchrony in top-down processing. *Nature Reviews Neuroscience, 2*, 704–716.

Engel, A. K., König, P., Kreiter, A. K., & Singer, W. (1991a). Interhemispheric synchronization of oscillatory neuronal responses in cat visual cortex. *Science, 252*, 1177–1179.

Engel, A. K., Kreiter, A. K., König, P., & Singer, W. (1991b) Synchronization of oscillatory neuronal responses between striate and extrastriate visual cortical areas of the cat. *Proceedings of the National Academy of Sciences USA, 88*, 6048–6052.

Engel, A. K., Roelfsema, P. R., Fries, P., Brecht, M., & Singer, W. (1997). Role of the temporal domain for response selection and perceptual binding. *Cerebral Cortex, 7*, 571–582.

Engel, A. K., & Singer, W. (2001). Temporal binding and the neural correlates of sensory awareness. *Trends in Cognitive Sciences, 5*, 16–25.

Felleman, D. J., & Van Essen, D. C. (1991). Distributed hierarchical processing in the primate cerebral cortex. *Cerebral Cortex, 1*, 1–47.

Fodor, J. A., & Pylyshyn, Z. W. (1988). Connectionism and cognitive architecture: A critical analysis. *Cognition, 28*, 3–71.

Fox, R., Todd, S., & Bettinger, L. A. (1975). Optokinetic nystagmus as an objective indicator of binocular rivalry. *Vision Research, 15*, 849–853.

Fries, P., Reynolds, J. H., Rorie, A. E., & Desimone, R. (2001). Modulation of oscillatory neuronal synchronization by selective visual attention. *Science, 291*, 1560–1563.

Fries, P., Roelfsema, P. R., Engel, A. K., König, P., & Singer, W. (1997). Synchronization of oscillatory responses in visual cortex correlates with perception in interocular rivalry. *Proceedings of the National Academy of Sciences USA, 94*, 12699–12704.

Fries, P., Schröder, J.-H., Roelfsema, P. R., Singer, W., & Engel, A. K. (2002). Oscillatory neuronal synchronization in primary visual cortex as a correlate of stimulus selection. *Journal of Neuroscience, 22*, 3739–3754.

Galambos, R., Makeig, S., & Talmachoff, P. J. (1981). A 40-Hz auditory potential recorded from the human scalp. *Proceedings of the National Academy of Sciences USA, 78*, 2643–2647.

Gray, C. M., König, P., Engel, A. K., & Singer, W. (1989). Oscillatory responses in cat visual cortex exhibit inter-columnar synchronization which reflects global stimulus properties. *Nature, 338*, 334–337.

Gray, C. M., & Singer, W. (1989). Stimulus-specific neuronal oscillations in orientation columns of cat visual cortex. *Proceedings of the National Academy of Sciences USA, 86*, 1698–1702.

Herrmann, C. S., Mecklinger, A., & Pfeifer, E. (1999). Gamma responses and ERPs in a visual classification task. *Clinical Neurophysiology, 110*, 636–642.

König, P., Engel, A. K., & Singer, W. (1996). Integrator or coincidence detector? The role of the cortical neuron revisited. *Trends in Neurosciences, 19*, 130–137.

Kreiter, A. K., & Singer, W. (1996). Stimulus-dependent synchronization of neuronal responses in the visual cortex of awake macaque monkey. *Journal of Neuroscience, 16*, 2381–2396.

Kristeva-Feige, R., Feige, B., Makeig, S., Ross, B., & Elbert, T. (1993). Oscillatory brain activity during a motor task. *Neuroreport, 4*, 1291–1294.

Laurent, G. (1996). Dynamical representation of odors by oscillating and evolving neural assemblies. *Trends in Neuroscience, 19*, 489–496.

Leopold, D. A., & Logothetis, N. K. (1996). Activity changes in early visual cortex reflect monkey's percepts during binocular rivalry. *Nature, 379*, 549–553.

Llinás, R., Ribary, U., Joliot, M., & Wand, X.-J. (1994). Content and context in temporal thalamocortical binding. In G. Buzsaki et al. (Eds.), *Temporal Coding in the Brain* (pp. 251–272). Berlin: Springer.

Logothetis, N. K., & Schall, J. D. (1989). Neuronal correlates of subjective visual perception. *Science, 245*, 761–763.

Metzinger, T. (1995). Faster than thought. Holism, homogeneity and temporal coding. In T. Metzinger (Ed.), *Conscious Experience* (pp. 425–461). Paderborn, Germany: Schöningh.

Müller, M. M., Junghöfer, M., Elbert, T., & Rockstroh, B. (1997). Visually induced gamma-band responses to coherent and incoherent motion: a replication study. *Neuroreport, 8*, 2575–2579.

Murthy, V. N., & Fetz, E. E. (1992). Coherent 25- to 35-Hz oscillations in the sensorimotor cortex of awake behaving monkeys. *Proceedings of the National Academy of Sciences USA, 89*, 5670–5674.

Neuenschwander, S., & Singer, W. (1996). Long-range synchronization of oscillatory light responses in the cat retina and lateral geniculate nucleus. *Nature, 379*, 728–733.

Pantev, C., Makeig, S., Hoke, M., Galambos, R., Hampson, S., & Gallen, C. (1991). Human auditory evoked gamma-band magnetic fields. *Proceedings of the National Academy of Sciences USA, 88*, 8996–9000.

Pulvermüller, F., Lutzenberger, W., Preissl, H., & Birbaumer, N. (1995). Spectral responses in the gamma-band: Physiological signs of higher cognitive processes? *Neuroreport, 6*, 2059–2064.

Rodriguez, E., George, N., Lachaux, J.-P., Martinerie, J., Renault, B., & Varela, F. J. (1999). Perception's shadow: Long-distance synchronization of human brain activity. *Nature, 397*, 430–433.

Roelfsema, P. R., Engel, A. K., König, P., & Singer, W. (1997). Visuomotor integration is associated with zero time-lag synchronization among cortical areas. *Nature, 385*, 157–161.

Roelfsema, P. R., König, P., Engel, A. K., Sireteanu, R., & Singer, W. (1994). Reduced synchronization in the visual cortex of cats with strabismic amblyopia. *European Journal of Neuroscience, 6*, 1645–1655.

Sheinberg, D. L., & Logothetis, N. K. (1997). The role of temporal cortical areas in perceptual organization. *Proceedings of the National Academy of Sciences USA, 94*, 3408–3413.

Shepard, R. N. (1990). *Mind Sights*. London: Palgrave.

Singer, W. (1999). Neuronal synchrony: a versatile code for the definition of relations? *Neuron, 24*, 49–65.

Singer, W., Engel, A. K., Kreiter, A. K., Munk, M. H. J., Neuenschwander, S., & Roelfsema, P. R. (1997). Neuronal assemblies: Necessity, significance, and detectability. *Trends in Cognitive Sciences, 1*, 252–261.

Singer, W., & Gray, C. M. (1995). Visual feature integration and the temporal correlation hypothesis. *Annual Review of Neuroscience, 18*, 555–586.
Steinmetz, P. N., Roy, A., Fitzgerald, J., Hsiao, S. S., Johnson, K. O., & Niebur, E. (2000). Attention modulates synchronized neuronal firing in primate somatosensory cortex. *Nature, 404*, 187–190.
Tallon-Baudry, C., & Bertrand, O. (1999). Oscillatory gamma activity in humans and its role in object representation. *Trends in Cognitive Sciences, 3*, 151–162.
Tallon-Baudry, C., Bertrand, O., Delpuech, C., & Pernier, J. (1997). Oscillatory γ-band (30–70 Hz) activity induced by a visual search task in humans. *Journal of Neuroscience, 17*, 722-734.
Vaadia, E., Haalman, I., Abeles, M., Bergman, H., Prut, Y., Slovin, H., & Aertsen, A. (1995). Dynamics of neuronal interactions in monkey cortex in relation to behavioural events. *Nature, 373*, 515–518.
von der Malsburg, C. (1981). *The Correlation Theory of Brain Function*. Internal Report 81–2. Göttingen: Max-Planck-Institute for Biophysical Chemistry. Reprinted (1994) in E. Domany, J. L. van Hemmen, & K. Schulten (Eds.), *Models of Neural Networks II* (pp. 95–119). Berlin: Springer.

Chapter 10: Hypothesized temporal and spatial code properties for a moment's working memory capacity: Brain wave "harmonies" and "four-color" topology of activated cortical areas

ROBERT B. GLASSMAN

Abstract

Working memory (WM), which operates under severe restrictions of time duration and item capacity (Baddeley, 1998), is the small doorway to all of cognitive growth and maintenance. What physical properties of brain matter are simple enough to be plausible foundations of a WM code at the "machine level," yet have sufficient possibilities of entering into informational patterns to be plausible foundations of WM at the cognitive level? Coherent EEG has temporal and spatial properties that seem suitable for binding together the cortical substrates of the codes for the attributes of a single cognitive item (Singer, 1994). Extending this principle, I suggest that there are reasons of informational economy why EEG frequency codes for up to three or four simultaneous items, that fill a moment's WM capacity, may be in harmonic relationships, within a single "octave"; i.e. in the small-integer ratios of "harmonies" (Glassman, 1999). There is an additional problem of how myriads of local activations of the cortical sheet (which stores codes for all of our long-term memories), with its extremely dense neuropil (Braitenberg & Schüz, 1998), may be allocated among the three or four items filling a moment's WM. Basic principles of plane topology and of graph theory suggest ways in which the cortical sheet may work within these limits. These principles include the "four-color" theorem for planar maps, a restriction to four convex figures for intersections of sets represented as Venn-diagrams, and the graph-theoretical principle that only up to four items may be comprehensively connected within a plane ("K4 is planar"; Glassman, 2000).

Small working memory capacity during each cognitive moment

Overview and general background

Intelligent creatures comprise myriads of structures and processes that change or grow just a little during each moment, as we flow with time. How big is a cognitive moment? What are the most significant psychological characteristics of a cognitive moment, and what do they look like in terms of brain function?

A famous 1956 review by George A. Miller coined the term "the magical number 7±2," concerning the puzzling persistence of this range in measurements of the numerousness of the momentary contents of consciousness. That paper is a focal point in the history of cognitive psychology. Its important summary of a great deal of prior research led to a larger body of subsequent research, in the ensuing decades, which concerned how many independent items (e.g., digits, letters, words or phrases, discrete ordinal positions in a rating scale judgment) may be borne in mind at once. In one part of his classic review, Miller referred to "immediate memory" as the psychological function concerned with this matter of capacity to hold only a small number of items in attention at once. Since that time, the terms "short-term memory" and "working memory" have come to be widely used, sometimes synonymously and sometimes making distinctions among different aspects, such as a subfunction devoted to storage as compared to ones devoted to executive functions, rehearsal, or intended actions (for reviews see Baddeley, 1998; Richardson, Engle, Hasher, Logie, Stoltzfus, & Zacks, 1996).

In addition, to *small capacity*, the other prominent defining characteristic of working memory (WM) is its *brief duration*. Consideration of this issue has evolved in different ways in different experimental contexts, often simply dichotomizing "short-term memory" versus "long-term memory." For example, in the animal neuroscience and experimental psychology literatures, "working memory" and "short-term memory" were often considered to cover a time span of as long as hours or even days (Glassman, 1999a; Glassman, Leniek, & Haegerich, 1998; Rosenzweig, 1996). Rosenzweig (1996) therefore suggested the label "intermediate-term memory" to help distinguish among phenomena that must surely involve different neural properties. Growing knowledge of neurophysiological functions, and of psychological phenomena having varied degrees of temporal persistence (e.g., see the excellent textbook by Rosenzweig, Breedlove, & Leiman, 2002), further suggests the provisional, simplifying nature of such terms as "working memory" and "long-term memory."

This paper emphasizes the immediate cognitive present. It therefore concerns both smaller estimates of WM capacity and smaller estimates of WM duration, on the order of a few seconds or less; however a peculiar paradoxical time-elasticity of WM duration is also mentioned.

The next subsection of this paper ("specific empirical background") comprises a very brief empirical review of our own findings concerning the generality across species of the small capacity of WM, for about 7±2 items. Some additional findings are reviewed that imply great "dimensional flexibility" of cognitive attributes, including time durations, which support the small constant WM capacity, both in humans and animals. This smallness of capacity may be critical to the function of working memory, in keeping its combinatorial processes in a fertile range—not too impoverished and not too overwhelmingly numerous and turbulent. After considering these issues, evidence is cited for an even smaller, core WM capacity. That is, during briefer moments, working memory capacity for *simultaneous,* vivid apprehension of a number of items is not as great as seven but is really only about three or four. This is particularly so if active efforts to rehearse the material are prevented.

This extremely modest quantitative range in cognition then becomes the focus in a quest to discover relevant "mid-scale" properties of the brain that suggest this same degree of numerousness. To that end, a set of examples is put forward of basic physico-mathematical phenomena that concern the range of 3–4 in elementary wave harmonics and in elementary graph theory and topology. The conjectures about how these phenomena may be relevant to brain waves and to cortical structure are intended both as plausible, testable hypotheses and as interesting prompts, which may elicit alternative hypotheses from other scientists, concerning the same basic mathematical principles. For example, although the ideas about harmonics, in this paper, emerged out of particular considerations of brain rhythms, it would be interesting if some of these properties of harmonics turned out to be consistent with some of the considerations about time quanta and rhythmic structure of perceptual and behavioral phenomena discussed by Geissler, Schebera and Kompass (1999; also Kompass, in press, and Geissler & Kompass, this volume, chapter 11).

Specific empirical background: Cross-species similarities

Radial maze findings with humans suggest a "universal constant" of WM. "The magical number 7±2" is our human WM capacity under a great variety of conditions, most often using verbal items. What is WM capacity in other species? Laboratory rats cannot be taught to speak, but it has been widely reported that rats in an 8-arm radial maze (see Figure 1) regularly attain a perfect score, by traveling down each arm of the maze once and only once to obtain the single food morsel at the end of each arm, even though their typical pattern for such foraging involves choosing arms in a random order. They do not seem to use odor trails. When the radial maze has 12 or 17 arms, rats' performance becomes imperfect. About a dozen years ago at Lake Forest College, we began a series of simple human behavioral experiments, in which we ask people to do analogous tasks, either with radial mazes drawn on paper or, on a nice summer day, walking along the arms of a 15 m diagonal radial maze painted on a large grassy area (see Figure 2). Our participants' average scores in 8-arm or 13- or 17-arm radial mazes turned out to be the same as those of lab rats (in these tasks, with good humor, we ask our human subjects to make it a fair contest with the rats by choosing arms in an unsystematic order; see O'Connor & Glassman, 1993; Glassman, Garvey, Elkins, Kasal, & Couillard, 1994; Glassman, Leniek, & Haegerich, 1998).

Figure 1: Illustration of a rat in a radial-arm maze.

Figure 2: Illustration of a person in the outdoor radial-arm maze analog task.

Duration paradoxes. The radial maze has interesting time properties as a "working memory" task, because it takes humans or rats some minutes to perform, much longer than the durations of typical verbal WM tasks. Moreover, during the 1980s some amazing results were reported, and replicated in several laboratories, for rats in a radial maze: If interrupted in the middle of a performance, they can complete it accurately even hours later (e.g. Beatty & Shavalia, 1980). We have also found that people in this spatial working memory task have a comparable ability, at least for delays of up to 15 minutes, filled with distraction (Glassman, Leniek, & Haegerich, 1998).[9] An extremely helpful anonymous reviewer for that paper offered specific references to the literature concerning the prodigious memories of bird species that cache food for the winter. Peculiarly, in laboratory tests with much shorter delays, such birds have been reported to reliably retrieve only about 4–8 hidden morsels (Bednekoff & Balda, 1996; Shettleworth & Krebs, 1986). For a comparable task with humans, we distributed 42 discretely marked hiding places in a large, grassy open field. People were asked to hide 12 place markers unsystematically in that situation, and then to retrieve the markers after five minutes of verbal distraction. Our subjects succeeded in retrieving an average of about seven of the 12 place-markers (see Figure 3; cf. Glassman, Knabe-Czerwionka, Branch, & Brown, 2001).

What might it mean that there is such a small constant range for WM capacity?

---

[9] We have also found the same WM capacity in a verbal task, having a formal similarity with the radial maze, in which subjects are required to recite a sequence of 8, 13, or 17 numbers or letters in random order, without repeating individual items. However, this task has less robust delay characteristics than does spatial memory in the radial maze. When we tried an 8-digit or 8-letter (A through H) random recitation task with an interference-filled 30-s delay interposed halfway, performance was significantly poorer than without the delay (average: 7 versus 7.7 items correct). It was impossible to do a standard digit-span (or letter-span) recall task, with 8 items, under such a delay condition.

Working memory capacity and personal history: Doing much with little

A narrow doorway. Each of us occupies only a tiny portion of space and grasps only a tiny cluster of things at any moment. Each life history is a tiny piece of world history. Indeed, in naturalistic approaches to theology, the term "Lord of History" is used in a rational way to address the proposition that evolution has naturally selected human beings to desire larger purposes than our individual knowledge suffices for - so we reach (Burhoe, 1975). At our best, in real life, each of us coordinates the data of past and present, progressing into the coming moment and wisely organizing intentions for the future. At such times, there is clarity, a sense of encompassing a great deal at once. However, when measured under controlled laboratory conditions, the capacity of WM to hold a number of independent items is quite small. The cognitive doorway of the present, remains narrow across diverse variations in the kinds of represented items.

Figure 3: Illustration of the outdoor cache-and-find task, in which 42 inverted flower pots served as discrete hiding places for items whose location had to be remembered by the participants

WM "theory of relativity". During every moment of life we stoop and squeeze through the WM doorway without realizing it. Human memory can be very patchy (Neisser & Hyman, 2000) and yet we manage to do well enough with it. The basic trick with WM, called "chunking," involves automatically organizing selected material from long-term memory into the few coherent items that can be held in attention at any given moment. Such "moments" are apparently structured by task-focus, in addition to absolute time duration limits. Thus, the time paradox of the strangely stretched durations in the radial maze procedures mentioned earlier—as well as other time paradoxes, including recency recall after slowly-paced list presentation, mentioned below—suggest a need for a "WM theory of relativity" (Glassman 1999a, 2000a). It looks as if a purposive thing is going on here—that our cognitive present "wants" about seven items to work with.[10]

Converging natural selection? A general question, here, is whether WM is crudely or finely engineered by natural selection. Are the quantitative properties of WM merely the outcomes of "cheaply opportunistic" natural selection processes that have settled into an evolutionary "groove"? It seems likely, instead, that WM capacity is a robust product that has "tuned in" to something deep in the logic of cognition, which concerns the fact that we are creatures living in a stream of time. In maintaining fairly constant capacity, WM may be supported by a variety of intermediate-time (ca. hours) memory effects, known by the names *priming, implicit memory, warm-up, procedural memory,* as well as *figure-ground* effects (Glassman, 1999a, pp. 480–484; Cowan, 1997).

But is "the magical number" seven or three? Under laboratory conditions, when you are pressed to keep the same 7±2 items in mind for several seconds, about half of the set is vivid while the remainder of this set undergoes active rehearsal. The notions of phonological loop and of sensory memory (Baddeley, 1998) are among the ways this fact has been conceptualized. More effortful tests of WM capacity also suggest that at any instant only about three or four items are strongly in focus, in WM. For example, the well-known textbook phenomenon of "recency" comprises about four items, each recalled with greater than 50% probability, at the end of a long list of presented words (Ashcraft, 2002, pp. 171–173; Neath, 1998, pp. 67–72). Interestingly, a "time paradox" has been discovered with recency. If a subject must try to recall the list after a 20-second delay that is filled with an interfering task, such recall of the most recent items is lost—but *not* if the *presentation* of words was at the slow pace of one word per 12 seconds, with interfering tasks occupying those 12-sec intervals (Bjork & Whitten, 1974, and reviewed in many textbooks). Other examples of effortful tasks yielding smaller WM capacity results include continuously tracking a sequence of input items occurring every second or every four seconds (Waugh & Norman, 1965); so-called "sentence span," "counting span," or "speaking span" WM tasks, which require a subject to keep items in mind while using them in sentences or calculations (Hitch & Towse, 1995; Lustig & Hasher, 2002; Miyake, 2001), and capacity for features of lines flashed at different orientations (Luck & Vogel, 1997). Recently, two students in my lab replicated Irwin Pollack's (1952) finding that when

[10] It might be interesting to systematically compare these "relativity" effects with the tendency of subjects, when explicitly judging time intervals, to be biased toward the comparison standard, whether it is longer or briefer (Eisler, 1996), or to compare them with other circumstances in which intended actions have a critical relationship with time judgments, as in driving (Michon, 1996).

the items to be remembered are musical tones played in a random sequence, immediate memory capacity is only about four tones. This seemed to be partly because, for people who are not experts in music, convenient symbols are not available to use in mental rehearsal (Glassman, McKenna, & Sienkiewicz, 2002). Findings supporting a conclusion that WM core capacity is only 3 or 4 items are reviewed in three earlier papers and an abstract (Glassman, 1999a, b; 2000a, b) and in a more recent extensive review (Cowan, 2001).[11]

Middle-scale phenomena. Reality is not a continuous blend. "Thinghood" occurs on many scales, but not at scales in between. There are galaxies, stars, and planets; also organisms, organs, cells, and molecules; etc., with big ontological gaps in between. Life occurs at a range of "middle scales" between elementary particles and the universe as a whole. These mid-scales are dense ontologically, so our scientific explorations of some levels may be insufficiently developed. Two sets of conjectures about temporal and spatial WM properties of the brain, in the following two major sections of this paper, illustrate this point. The first concerns a hypothesis that particular EEG harmonic phenomena underlie the 3s or 4s of WM capacity, at intermediate scales between large cortical masses and single neuron activity. The second suggests that certain basic principles of plane topology also underlie a core capacity of WM for handling 3 and 4 independent items in the functioning of the sheet-like cortex. These two hypotheses attempt to integrate knowledge, at the neural level, both of "time and mind" and of "space and mind."

EEG harmonics: Hypotheses about *time properties* of WM brain signals

Which aspects of brain waves might comprise cognitive signals? Diffuse EEG activity is well known to correlate reliably with states of arousal, attention, and certain other cognitive phenomena, yet it is not known to what degree such activity actually plays a causal, connective role, as opposed to being a byproduct. Now being widely investigated is the hypothesis that the neural representations of attributes undergo *binding*, into unified cognitions, by means of synchronous or phase-locked, coherent relations among spatially distributed neural systems. Such *electrical coherence* potentially addresses the causal-linkage questions of mediation between short- and long-term memory, and of how dispersed brain regions, which individually tune into the diverse properties of a psychological event might temporarily configure and reconfigure in each succeeding moment, to achieve *psychological coherence* (e.g., Kavanau, 2002; Singer, 1993; Schack, Vath, Petsche, Geissler & Möller, 2002; see also Engel, this volume, chapter 9). However, because WM contains more than a single item at a time, a single set of coherent binding relationships seems insufficient.

[11] I thank both Simon Grondin and Richard Block for independently suggesting this interesting paper, at the Time & Mind 02 conference.

Two WM layers. According to the binding hypothesis a single perceived object or single concept ("conceit") has its attributes tied together, during its moment in attention, by means of phase-locked activity at a given frequency among the brain systems that it evokes. If this hypothesis about a *single cognized object* is valid for diverse modes of cognition, then it entails the additional problem of how the brain can act as a substrate for the *several conceits* that are in WM at any moment. The terms "*single-element binding*" and "*small-cluster binding*" might be coined to describe two "layers" of WM. In this paper the principle of two WM layers is proposed only in general terms of associative interplay of chunks, without undertaking the additional matter of syntactic structuring (Shastri, 2002).

Might EEG signaling properties play a causal cognitive role in both layers of WM? Candidate phenomena must be appropriately differentiated within time and brain space. Time durations of the briefest cognitive phenomena, fractions of a second, comprise a rough clue to the duration within which candidate EEG phenomena should exhibit fairly stable characteristics. These hypothetical wave phenomena might be obvious in raw EEG recordings or might require "mining" of the data. Brain wave harmonics are hypothesized here as candidates having appropriate properties.

Is "harmonious" cognition supported by literal harmonies in EEG signaling?

A temporal property of waves in harmony. Waveforms, in any medium, can be analyzed as summations of component sinusoidal waves, but certain wave relationships are special. The human ear, across diverse cultures, is pleased by simple musical harmonies. This fact correlates with a certain "efficiency" in the signaling properties of harmonies: Waves whose frequencies are in the low-integer ratios of harmonies (especially 3/2, 4/3) have summed waveforms that have the *shortest possible wavelength* (see Figure 4). Harmonious frequencies can also occur in fast triplets; for example the major triad (ratios 3/2 plus 5/4 present at once) has a wavelength only four times that of the fundamental frequency (Glassman, 1999b, 2000a). This might be important if EEG has a role as a vehicle for signals at the neural level, to serve working memory at the level of cognition. Working memories often have to fade within a few seconds when there is a continuing flow of information. Our ability to ignore, shelve, or discard information at an appropriate rate is as important as our ability to acquire information (Sachs, 1967).[12] Studies of adults and of developing children, have shown that both the capacity of working memory and intellectual capacity vary directly as a function of the speed of WM processing (see reviews in Cowan, 1997; Weinert & Schneider, 1995; Weiss, 1992; also the excellent developmental psychology textbook by Berk, 2003, pp. 274–277).[13] Pursuing the EEG harmonics hypothesis, if three items in WM were coded in a gamma EEG band "major

[12] I thank Bjorn Merker for this interesting source.

[13] However, the role of EEG frequency in intelligence remains controversial (see Andreassi, 2000, pp. 41-60 for review).

triad," say, of 40 Hz, 50 Hz, and 60 Hz, a single summed wave would take 0.1 sec. At least two waves are necessary to establish that there is a consistent signal, and the two-tenths of a second that this implies is, appropriately, at the brief end of the range of meaningful cognitive time intervals (for example, it is a simple reaction time).

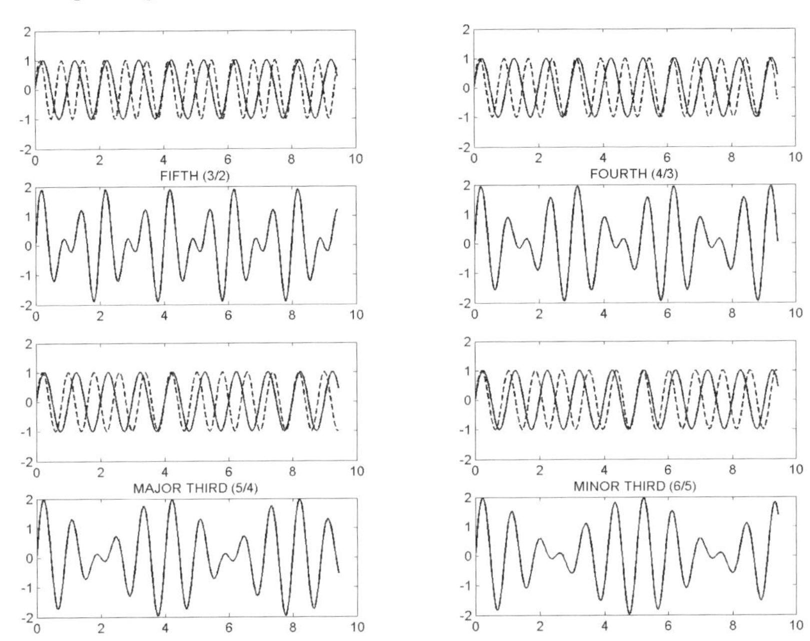

Figure 4: Waves that are in harmony with each other have short wavelengths in their moment-by-moment summed waveforms. For example, with the harmony that is classically considered best, the "perfect fifth," there is a 3:2 ratio between wavelengths, and the summed waveform has only twice the period of either individual contributing wave. In this figure, the upper graph for each illustrated harmonic ratio shows the two contributing waves, while the lower graph shows their sum. Note that the peaks of the sum occur where the two contributing waves are closest to having synchronized peaks. Since these graphs use the sine function, perfect synchrony occurs only at the zero-crossing. The peaks would synchronize if it were a cosine function. The formula for the major third is $y = \sin[2\Pi x] + \sin[(1.25)(2\Pi x)]$. Plotted using MATLAB 6.5 software.

Adequate differentiation. But how plausible is it that a sinusoidal EEG wave, as simple and low in frequency as such waves are, could be the brain-activity aspect of something as complex as a cognition in WM? Because WM capacity is so small, at any given moment there may be a correspondingly small demand for signal differentiation in the time or frequency domain binding frequencies. As for the myriads of features in long-term memory, from which a selection is tapped by WM in each moment, this differentiation must reside largely in the exquisite, dense spatial structure of the brain.

Useful properties of sinusoids. Brain wave components that are basically sinusoidal might have three advantageous properties as signals:

(1) *Form constancy with phase jitter.* Sinusoidal form constancy of summated sinusoids is sustained in the face of time-of arrival (or phase) errors. Two sinusoidal waves of a given frequency always summate to a sinusoid of that same frequency.

(2) *Reliability.* Rhythmic repetition of a signal is often part of reliable communication. That fact, together with Item 1, above, implies that temporal extension of a signal allows coordination of multiple converging signals in spite of small time-of-arrival mismatches.

(3) *Form constancy with differentiation or integration.* Because the derivative and integral of a sinusoid are also sinusoids, any given brain system can use a constant form of signal *recursively*, while registering an input's rate of increase or decrease (derivative) or, conversely, an input's cumulative signal strength over some time interval (integral).

Thus, it is conceivable that at any instant up to three or four different EEG frequencies could act as markers of the up to three or four items that are then most vivid in working memory. Such a pattern of brain wave harmonics would have to encompass all the regions of brain tissue whose feature-analyzing properties lead them to be called upon for those particular cognitive objects. These ideas are reviewed in more detail in two papers (Glassman, 1999b, 2000a)

Other suggestions of threeness. Using primarily a different set of justifications, than are presented here, Shaw, Bodner, and colleagues have argued that there is a deep relationship between music and brain function.[14] They argue also that threeness is important, and coin the term "trion" in a hypothesis about the elementary unit of neural information processing (Shaw, 2000). An additional justification of their emphasis on threes, or alternatively my own hypotheses about threes, is in a combinatorial fact long known in computer science. This is that if it were feasible to replace the typical "yes-no," "on-off," or "1 versus 0" *binary* most-fundamental structural elements of computers with *trinary* elements (each having a range of 3 activation levels), this would allow a more efficient simple representational system. That is, fewer elements plus states of those elements would be needed in order to represent, label, or count a number of objects. This is because a base-3 counting system is more efficient than our decimal system, a binary system, or any other integer-base (Hayes, 2001). Such an efficiency becomes meaningful when dealing with the extremely large numbers of elements in a complex information-processing "machine" (com-

---

[14] It is also appropriate to cite an interesting compendium of other ideas on "biomusicology," concerning deeper significances of music in evolution of human behavior and brain (Wallin, Merker, & Brown, 2000).

puter or brain) together with the extremely large number of things to be represented in a knowledge system.[15]

Octaves: Possible relevance of a sampling frequency rule in signal processing engineering

Nyquist frequency. A hypothesized *octave band restriction* on brain wave signaling is part of the foregoing hypothesis about a cluster of 3 or 4 EEG harmonics. Although EEG signals are continuous, the octave band hypothesis seems to be related to the following basic principle of digital signal processing, called the *sampling theorem*: A wave must be sampled at a frequency that is at least twice as high as the highest frequency in the signal (called the *Nyquist frequency*). If the sampling rate is lower than that, spurious frequencies, called *aliases*, are introduced that appear in the same band as the frequencies of interest. For example, if we wish to digitize EEG waves from 0 to 100 Hz we will need a sampling frequency of at least 200 Hz (Lyons, 2001; Chugani, Samant, & Cerna, 1998).

The phenomenon of *aliasing* seems to be related to that of *difference frequencies*, which occur when two continuous-wave frequencies are present in the same medium. One form of these comprises *beat frequencies*, when two frequencies are close. For example, if two musical instruments are in slightly different tune, with middle C of one of them set at 261 Hz while middle C of the other is 262 Hz, when both are played at once, the sound will wax and wane once a second (262–261 = 1).

Octaves prevent being beaten by beats. Reminiscent of the Nyquist frequency, a basic algebraic principle implies that so long as two frequencies are within a single octave—that is, so long as the higher frequency is less than twice the lower frequency—no difference rhythm can occur within the same octave. This is simply to say that if you take two numbers, x and $2x$, the difference $2x-x$ cannot fall between x and $2x$. This suggests that if sinusoids act as simple signals in the EEG, they can engage in combinatorial play without introducing spurious, distracting aliases, so long as the three or four signal frequencies that represent WM chunks remain within a single octave. One possible EEG range that would fulfill this requirement would be a gamma band that extended, say, from 40 to 80 Hz.[16] However, in beginning to

---

[15] To illustrate with approximations using small numbers, consider that two digits of our base-10 system can count from 0 to 99. With a binary system it takes between 6 and 7 digits to count or label the same 100 items ($2^6 = 64$, $2^7 = 128$; and $100 \approx 2^{6.644}$). And with a trinary system it takes between 4 and 5 elements ($3^4 = 81$, $3^5 = 243$; and $100 \approx 3^{4.192}$). The "efficiency" arises from the fact that two decimal elements have a total of 10 + 10 = 20 levels; the total number of required levels for binary and trinary counters are, respectively, 13.288 (= 2 × 6.644) and 12.576 (= 3 × 4.192). Only the non-integer base e = 2.718... is more efficient than the integer 3 as a base.

[16] A variety of additional musical phenomena, including the "fundamental bass," difference rhythm "roughness," and the psychological need for resolution of chord patterns (Pierce, 1992; Plomp, 1976; Rossing, 1990) are also suggestive of possible brain wave counterparts of WM information processing (Glassman, 1999b).

explore this issue empirically in my lab, we are considering the full range of EEG frequencies that we are able to record, an octave at a time.

Topology of active cortical patches: Hypotheses about *spatial properties* of WM signals

Although it is possible to develop arguments about *spatial* periodicities (in one or more dimensions) that are analogous to the foregoing arguments concerning temporal periodicities, another form of argument, using basic principles of graph theory and topology applied to the massive human cortex, may more interestingly fit the same constraint of WM-item threeness or fourness as is suggested by the psychological data and by considering the possibility of EEG harmonics within an octave. This section of the paper discusses the intriguing near-two-dimensionality of cortex, and then explains how 3s or 4s might come out of that gross anatomical circumstance.

Interesting peculiarities of cortical shape and architecture

The approximately 90 per cent of the human brain that is our cortex must contain virtually all of the immensity of information in long-term memory, which is drawn upon by the WM system, with its capacity to mobilize only a few chunks during any one moment. Strikingly, the cortex is a sheet, in humans only slightly thicker than in the mouse. Mountcastle's (1997) estimate of 2,600 cm^2 implies— amusingly—that if the cortical sheet had a square perimeter it would have almost exactly twice the linear dimensions of a typical graduation "mortarboard" cap (and see Hofman, 1988, for implied larger estimated area). Indeed, the mammalian cortical thickness of 2 to 3 mm is the same as the thickness of such a cap. These figures suggest the whimsical, species-centric or "human chauvinist" remark that human beings are "the college graduates of biological evolution" (see Figure 5).

The amazing cortical sheet comprises tens of billions of pyramidal neurons, so densely packed that there are 100,000 of them beneath each square millimeter of its outer surface (Braitenberg & Schüz, 1998). Each cubic millimeter contains about a billion synapses, distributed over approximately 1–2 kilometers of axons and 456 meters of dendrites (Braitenberg & Schüz, 1998; Abeles, 1991; DeFelipe, Marco, Busteria, & Merchán-Pérez, 1999). Deliberately mixing metric and English units to describe these awe-inspiring facts about the cortex leads to the evocative alliteration "miles within millimeters" (Glassman, 2002).

Additional neuroanatomical evidence of sheet-like functionality. Although the cortex comprises about six layers there is a unifying "vertical" organization spanning from the outer surface of the cortex through its full 2–3 mm depth. The neurons are generally clustered as "cortical columns," typically about 300 to 600 µm in diameter, having much greater interconnectedness within than between columns (Braitenberg & Schüz, 1998; Mountcastle, 1997, p. 702; White, 1989, pp. 8–12, 20–29). This anatomical unity in the dimension of its thickness again implies that the cortex functionally has a sheetlike character.

Figure 5: Humorous illustration of what it would look like if a graduation cap were the area of the human cortex (picture modified from the Microsoft Word clip art set). Such a cap would have about twice the linear dimensions of, but exactly the same thickness (2–3 mm) as, a typical graduation cap.

Moreover, most of the commerce of the cortex, by far, is "private," an internal matter. Even in primary sensory cortex, no more than 20 percent of the synapses receive information from outside the cortex. Although 80 per cent of cortical neurons contribute axons to the white matter, overall, more than 98 percent of cortical white matter comprises connections not to subcortical areas or to the spinal cord and the body, but from one cortical area to another (White, 1986, cited by Braitenberg & Schüz, 1998, p. 43). Again, this implies a sheet-like quality of the cortex's woven fabric.

The cortex as a metaphorical "membrane." One of the natural selection pressures toward the thinness of the "vertical" dimension of the cortex may concern the severe cognitive constraint of small WM capacity. Before proceeding with this idea it is necessary to acknowledge that the severe anatomical constraint of a thin cortex may primarily be the result of many other factors. For example, it may be a necessary concomitant of ontogenetic organizing processes or of other aspects of function in a three-dimensional universe. Indeed, our bodies have *many* sheet-like structures, such as cell membranes and blood vessel walls. These have a variety of functions, including separation, confinement, selective transport, and uniformly ready availability for signals. Some of these functions, indeed, are metaphorically suggestive of a role of cortex as a "wall" (with a "narrow doorway" of WM) or "membrane" whose limited WM "permeability" divides a living, intelligent individual's past from his or her future.

Combinatorial considerations

Creative order lives in a sheet, between chaos and inert stability. Returning to the "why" issue discussed earlier, perhaps the anatomical and functional features and constraints attending the sheet-like cortex yield a good compromise between combinatorial explosiveness on the one hand and fecundity in the combinatorial play of associations on the other. The design of brain must make it convenient for mind to steer between the Scylla and Charybdis of dull inertness versus sheer chaos (Glassman, 1999a). Related issues of degrees of retention versus forgetting have been discussed by others, at the levels of collective memory of a society (Nora, 1996; Richelle, 1996, p. 12) or the pace of life in different cultures (Helfrich, 1996; Levine, 1996). These matters seem related also to the issue of surface details versus meanings (and "gists") understood by individuals using language (Zangwill, 1972; Ericsson & Kintsch, 1995; Engle & Conway, 1998; Libby & Neisser, 2001), and the issue of cortical connectivity (Merker, in press).

Suppose too much. A brief illustration of this matter of combinatorial explosiveness may be used in a suggestive way in regard to WM capacity of 3, while by no means proving the point nor yet tying it clearly to cortical shape. Consider that with, say, N = 8 cognitive items independently active at once, carrying out a hypothetical exhaustive search over all possible combinations would require $2^N = 2^8 = 256$ separate tests. However, exhaustively considering all combinations of only three items requires only $2^3 = 8$ tests. Pursuing this hypothetical point, eight sequential tests, each involving a major triad of gamma EEG frequencies, could take place as quickly as in 1.6 s ($= 0.2 \times 8$), a behaviorally realistic decision time, and a time within the 1.5–2 seconds range that has been considered the decay-time limit of a vivid WM trace unaided by rehearsal (see reviews by Baddeley, 1998, pp. 52–63; Cowan, 1997, pp. 172–178).

Suppose too meager. Coming at this same quantitative issue "from below," consider that with only a single item in mind at one time, things are extremely dull; no associations are possible. If two items are active at once, that permits associative activity, but only of a meager sort—a choice between linking the items and letting the opportunity pass. But life is highly contingent. Whether a particular relationship holds depends, in general, on *some third factor* or, perhaps, WM chunk of factors. This consideration may come close to expressing the raison d'être of WM, as well as the fundamental, quantitative "design problem" for WM capacity.[17]

Connectivity: Density, sparseness, and degrees of separation in graph theory

Before proceeding to consider specifically how threes or fours might follow from the sheet-like nature of cortex, it is appropriate to briefly consider the general issue of emergence of middle-scale properties from connection density factors at the microscopic level. In spite of the extreme density of connectivity in the cortex, there is only approximately a chance in a thousand that any one pyramidal neuron within a small cortical region will synapse on any of the others within that region (Braitenberg & Schüz, 1998). From a statistical point of view, the local connectivity seems random; it looks as if the synapses "rained" down (Braitenberg & Schüz, 1998, p. 51). Thus, the extreme denseness of neural fibers ("miles within millimeters") should

[17] These considerations are somewhat akin to a theory that a WM capacity of seven optimizes the rapid detection of statistical correlations (Kareev, 2000; I thank Nancy Brekke for pointing out this source.)

not distract us from a countermanding sparseness of local connectivity among the basal dendrites of neighboring pyramidal neurons.

Readiness for networking. In graph theory, when extremely large numbers of elements are sparsely interconnected in a random manner (rather than in a regular manner, or "lattice"), a condition readily develops in which the individual nodes have surprisingly few "degrees of separation." That is, there is a path from one node to any other node via only a few intermediate nodes (Watts & Strogatz, 1998; Watts, 1999; Hayes, 2000). Threshold discontinuities comprise a related aspect of random graphs. Thus, the sparse, random local connectivity among pyramidal neurons within a cluster may comprise a pregnant situation, a sort of arena of productive freedom, in which cohesive dynamic patterns of neural activity are always on the threshold of creation as a result of small adjustments in the connection weights at a few synapses, while always ready to dissipate and give way to alternative patterns. One may imagine these sorts of activity fluctuations dynamically unifying the cortex as a whole, or unifying large or small subareas of cortex rather independently of each other, each in its own tightly-coupled neural activity circuits, for the duration of a "WM moment." In graph theory, regions of complete internal coupling, which are loosely coupled to other such regions, within which there are also few degrees of separation, are known as "cliques" (Watts, 1999, pp. 37, 102–109). Thus, in the present context we may think of a "clique," hypothetically, as the graph-theoretical aspect of single-element binding of individual WM chunks.

Freedom to associate. Considering the other "layer" of WM function, small-cluster binding of threes or fours of items must also entail a kind of freedom. When we participate in laboratory experiments, WM contents are stilted and predetermined. In real life the brain takes things as they come. Even during deliberate thinking, in solitude, thoughts emerge from preceding thoughts rather than being rehearsed. Therefore, there has to be arbitrary freedom of associations among all the chunks in WM at any moment. Thus, again, freedom of association must occur on both levels of binding. First, the brain's representational substrates for *attributes* of each chunk must be readily available to be evoked in any combination for binding together into a unified percept or concept and quickly unbinding in readiness for what comes next. Second, given the independent coalescence of up to three or four of such cognitive objects at once, all three or four must be free to associate in any combination.

Three principles, constraining to threes or fours, in topology and graph theory

Three additional basic principles from mathematical graph theory and topology imply that in a planar environment information processing may be restricted to three or four independently varying items.

(1) The four-color principle and shared "subpatch" boundaries. The near two-dimensionality of the cortex suggests considering its functional regions as *patches*, which are available for activation. If that makes sense, then the famous four-color principle of topology (Barr, 1989; Saaty & Kainen, 1986) suggests that a region that is committed to WM during some short time interval may be the scene of combinatorial play among up to four "*subpatches*." Such combinatorial play involves both

competition of psychological attributes for appropriately tuned cortical substrates and also associative interactions among WM chunks.

Patchiness of brain regions has long been considered in describing anatomical connectivity over the long term. A much more dynamic, short-term cortical patchiness may be relevant to WM capacity. It seems worthwhile to consider this possibility apart from the issue of hypothesizing any particular, typical size of such patches. Conceivably, even a cortical region as small as a 0.5 mm cortical column might be considered as a flat "patch," although such columns are taller than they are wide. This is because columns may well have much more rigorously homogeneous feature-analyzing properties in the "vertical" dimension than horizontally, where the sparse, apparently random local connectivity among basal dendrites implies looser horizontal coupling.

Element binding of a single conceit must invoke myriads of widely distributed cortical columns. Cognitive associative activity *among* the pairs, triplets, or quadruplets of conceits held at one time in WM (the small-cluster level of binding) *may* also always involve long-distance relations as its primary feature. However, there are interesting implications if, instead, associative activity occurs largely by virtue of *local* interactions within each patch, while activity in the long-distance cortico-cortical axons is, hypothetically, restricted to binding *within* each chunk. This could happen if some significant number of the patches of activated cortical tissue *each* momentarily commit to *all* of the chunks in WM. For the duration of that WM moment, the subpatch that represents a given chunk, in each of these patches, is bound to its respective counterpart in every other patch. Such a thing might happen most readily in the fronto-limbic areas in which there is an extreme degree of convergence from all other areas and which much evidence shows to be involved in WM (Merker, in press). Each such hypothesized patch then might be divided up dynamically into as many parts as there are items in WM. If each such subpatch has a well-defined boundary, then the cortical embodiment of a momentary association among simultaneously active WM chunks is in a dynamic interplay occurring at the *shared boundaries* of the subpatches that represent those chunks.

These hypothetical working memory patches and subpatches would have their brief moments of existence as dynamically changing amoeboid shapes, which are continually moving around each other and also denting each other with new protuberances, while maintaining edge adjacencies. Again, because of global binding *within* each WM chunk, there must be closely correlated activity *among* all the subpatches, distributed over the cortex, which represent a given chunk. Topologically, mutual shared boundaries occur readily when an area is divided three ways, and only up to four "subpatches" can simultaneously freely associate, each with every other, in this way. It is impossible to add a fifth closed region in such a way that each of the regions shares a border with every other. This hypothesis about cortical subpatch associations is an apparent corollary of the famous four-color principle for geographic maps. That principle says that no more than four distinguishing markers (e.g., colors) are needed for designating subareas of any map that can be drawn on a plane, in such a way that no two regions having the same marker share a boundary (see Figure 6).

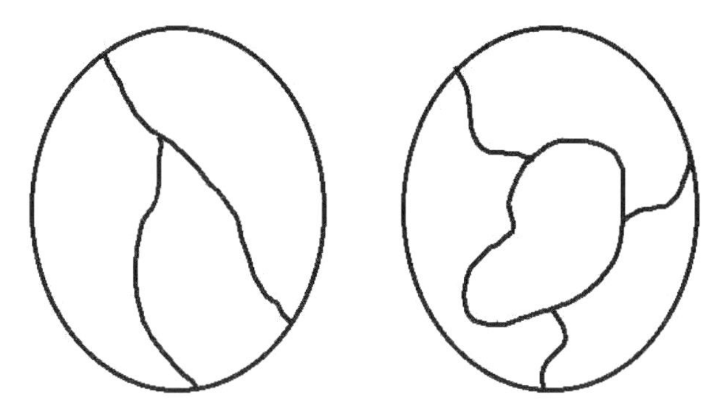

Figure 6: Up to four discrete subareas of a plane each can make mutual contact with every other subarea.

(2) Connecting up to four nodes without crossings. A basic principle of graph theory leads to the same conclusion about a limit of four, for contacts among freely associating local regions within a plane. This theorem is that "K_4 is planar" (see Figure 7), while "K_5 is nonplanar".[18] That is, all six possible connections can be drawn among four nodes, without any crossings of lines. However, if a fifth node is added, then at least two of the connecting lines must cross each other in the plane. Of course, even with a mere 2–3 mm third dimension of depth, there is more than ample room in the cortex for slender axons to cross each other in order to yield differentiation of associative activity on a scale of tens of microns. However, this paper is pursuing the premise that the cortex is, to a large degree, functionally a plane, in which there is cognitively significant cortical activity that requires unitary commitments on a larger scale of cortical volumes, perhaps a scale of approximately 0.1 mm horizontally, and extending through one or more cortical layers vertically (see Krahe, Kreiman, Gabbiani, Koch, & Metzner, 2002, for consideration of correlated activity in overlapping pyramidal neurons, using a vertebrate model system). Within that scale, the dendritic trees of thousands of pyramidal neurons overlap. If such is the case, then avoiding interfering crosstalk among WM chunks might require the K_4 restriction. It is also possible to envision an intermediate scale of connectivity in which much of the cortical columnar depth is unitarily committed in each subpatch, but an additional association might skirt these entrenched chunks via "overpass connectivity" employing the wide lateral spread of apical dendrites in the cortical molecular layer.

[18] "K" stands for "complete graph," meaning every node is connected to every other.

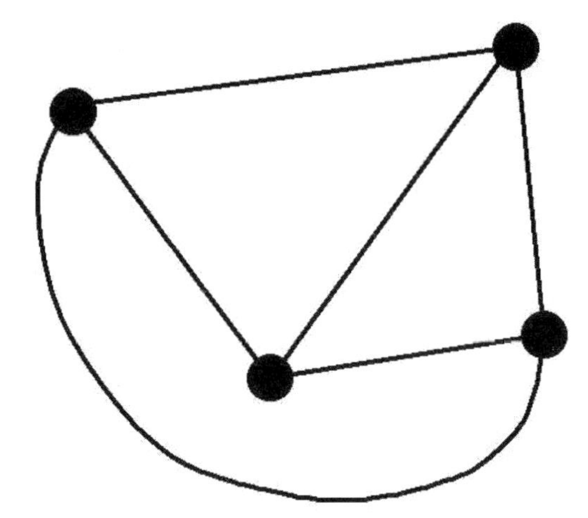

Figure 7: Up to four nodes can be connected within a plane without any crossing of connections.

(3) Only up to four convex Venn figures can intersect exhaustively. The topological considerations discussed above concerned edge-contacts as the hypothesized mode of association among discretely bounded subpatches. Coincidentally, the same inference about a limit of four subpatches is suggested by a very different topological premise, under which the hypothetical subpatches that embody WM chunks must overlap, or interpenetrate, in order to achieve associations. Such embodiments of cognitive items can be considered as if they were like the Venn diagrams of symbolic logic. Three convex figures (having no inward bends; Yaglom & Boltyanskii, 1961) easily achieve a comprehensive set of intersections, while up to four convex figures can intersect exhaustively, with some necessary stretching and squaring off (see Figure 8). Beyond that number, serpentine shapes are required (see Figure 9). While such diagrams with an exhaustive set of intersections can be achieved, with effort, by a whole, intelligent person drawing intersecting figures, it seems more plausible that the level of "intelligence" of associative play among cortical subpatches must have a more primitive quality, which involves simpler, amoeba-like dynamics of random exploratory expansions, contractions and translations of convex boundaries. Greater numbers of items interacting in this way entail greater possibilities of errors (either omissions or repeats).

These arguments, that applications of plane topology and of graph theory to the cortical sheet are deeply consistent with the cognitive restriction of core WM capacity to three or four items, are developed more fully elsewhere, in a paper that also discusses some economies of time and metabolic expense that would be achieved if the main neurophysiological work, as WM chunks undergo combinatorial associative play, indeed occurs at the level of local interactions, while long-distance cortico-cortical activity plays a reduced role, limited to global binding *within* individual chunks (Glassman, in press).

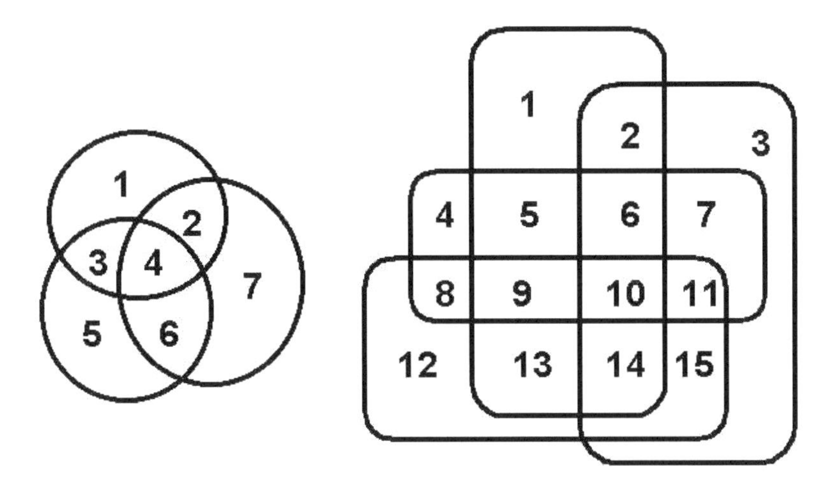

Figure 8: Up to four convex figures can be made to exhaustively intersect, in the manner of Venn diagrams (drawn after Barr, 1989).

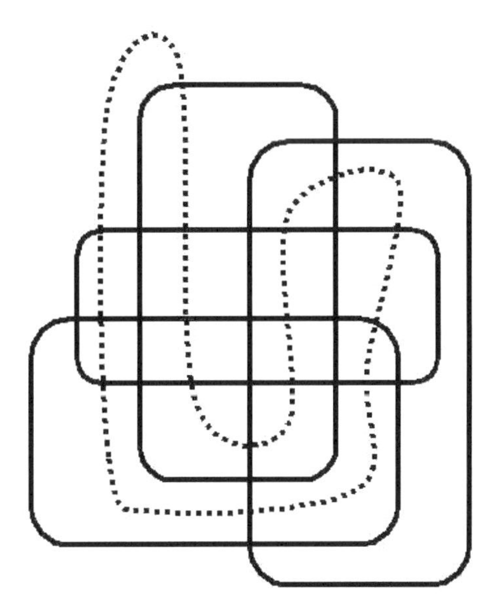

Figure 9: A fifth Venn figure, which exhaustively intersects each of the others, must be concave.

Concluding comment

I hope these efforts, to bring together threes or fours of WM with threes or fours of the mathematics of waves and of surfaces, comprise plausible conjectures about brain structure and function, and that they deserve a role as a set of lenses for viewing empirical data. In doing relevant empirical work, it may make sense to be exploratory rather than strictly hypothetico-deductive. In my undergraduate teaching laboratory, we are working on a computer program to seek relationships among the peaks of our own brain wave spectra during WM tasks.

References

Abeles, M. (1991). *Corticonics*. Cambridge: Cambridge University Press.
Andreassi, J. L. (2000). Psychophysiology. Fourth edition. Mahwah, NJ: Erlbaum.
Ashcraft, M. H. (2002). *Cognition.* Third edition. Upper Saddle River, NJ: Prentice-Hall.
Baddeley, A. (1998) *Human memory: Theory and practice*. Revised edition. Boston: Allyn and Bacon.
Barr, S. (1989) *Experiments in topology*. New York: Dover.
Beatty, W. W., & Shavalia, D. A. (1980). Rat spatial memory: Resistance to retroactive interference at long retention intervals. *Animal Learning and Behavior, 8*, 550–552.
Bednekoff, P. A., & Balda, R. P. (1996). Observational spatial memory in Clark's nutcrackers and Mexican Jays. *Animal Behavior, 52*, 833–839.
Berk, L. E. (2003). *Child development*. Boston: Allyn & Bacon.
Bjork, R. A., & Whitten, W. B. (1974). Recency-sensitive retrieval processes in long-term free recall. *Cognitive Psychology, 6*, 173–189.
Braitenberg, V., & Schüz, A. (1998). *Cortex: Statistics and geometry of neuronal connectivity*. Second edition. Berlin: Springer.
Burhoe, R. W. (1975). The human prospect and the "Lord of History." *Zygon, Journal of Religion and Science, 10*, 299–375.
Chugani, M. L., Samant, A. R., & Cerna, M. (1998). *LabVIEW signal processing*. Upper Saddle River, NJ: Prentice-Hall.
Cowan, N. (Ed.), (1997). *The development of memory in childhood.* Hove, East Sussex, UK: Psychology Press.
Cowan, N. (2001). The magical number 4 in short-term memory: A reconsideration of mental storage capacity. *Behavioral and Brain Sciences, 24,* 87–185.
DeFelipe, J., Marco, P., Busturia, I., & Merchán-Pérez, A. (1999). Estimation of the number of synapses in the cerebral cortex: Methodological considerations. *Cerebral Cortex, 9,* 722–732.
Eisler, H. (1996) Time perception from a psychophysicist's perspective. In H. Helfrich (Ed.), *Time and mind* (pp. 65–86). Seattle, WA: Hogrefe & Huber Publishers.
Engle, R. W., & Conway, A. R. A. (1998). Working memory and comprehension. In R. H. Logie & K. J. Gilhooly (Eds.), *Working memory and thinking* (pp. 67–91). Hove, East Sussex, UK: Psychology Press/Taylor & Francis.
Ericsson, K. A., & Kintsch, W. (1995). Long-term working memory. *Psychological Review, 102*, 211–245.
Geissler, H-G., Schebera, F-U, & Kompass, R. (1999). Ultra-precise quantal timing: Evidence from simultaneity thresholds in long-range apparent movement. *Perception and Psychophysics, 61,* 707–726.

Glassman, R. B. (1999a). A working memory "theory of relativity": Elasticity over temporal, spatial, and modality ranges conserves 7±2 item capacity in radial maze, verbal tasks and other cognition. *Brain Research Bulletin, 48,* 475–489.

Glassman, R. B. (1999b). Hypothesized neural dynamics of working memory: Several chunks might be marked simultaneously by harmonic frequencies within an octave band of brain waves. *Brain Research Bulletin, 50,* 77–93.

Glassman, R. B. (2000a). A "theory of relativity" for cognitive elasticity of time and modality dimensions supporting constant working memory capacity: Involvement of harmonics among ultradian clocks? *Progress in Neuro-Psychopharmacology and Biological Psychiatry, 24,* 163–182.

Glassman, R. B. (2000b). Cortical plane topology of working memory capacity: Four-chunk limit whether attribute-representation subpatches overlap in, or tile, patches of cortex. *Society for Neuroscience Abstracts*, Volume 26, Part 1, p. 706, #263.11 (November, 2000, New Orleans).

Glassman, R. B. (2002). "Miles within millimeters" and other awe-inspiring facts about our "mortarboard" human cortex. *Zygon: Journal of Religion and Science, 37,* 255–277.

Glassman, R. B. (in press). Topology and graph theory applied to cortical anatomy may help explain working memory capacity for three or four simultaneous items. *Brain Research Bulletin.*

Glassman, R. B., Garvey, K. J., Elkins, K. M., Kasal, K. L., & Couillard, N. (1994). Spatial working memory score of humans in a large radial maze, similar to published score of rats, implies capacity close to the magical number 7±2. *Brain Research Bulletin, 34,* 151–159.

Glassman, R. B., Knabe-Czerwionka, J. J., Branch, A. M., & Brown, P. B. (2001). Proud to be a bird brain: Human working memory for caching and finding items in an outdoor area as good as birds in a laboratory. *Society for Neuroscience Abstracts, 27,* #74.4.

Glassman, R. B., Leniek, K. M., & Haegerich, T. M. (1998). Human working memory capacity is 7±2 in a radial maze with distracting interruption: Possible implication for neural mechanisms of declarative and implicit long-term memory. *Brain Research Bulletin, 47,* 249–256.

Glassman, R. B., McKenna, A. K., & Sienkiewicz, A. P. (2002). Extension of a classic study showing working memory capacity for random musical tones is only four: Effect of frequency range. (Teaching of Neuroscience poster) Program No. 22.69. *Abstract Viewer/Itinerary Planner.* Washington, DC: Society for Neuroscience. CD-ROM

Hayes, B. (2000). Graph theory in practice: Part II. *American Scientist, 88,* 104–109.

Hayes, B. (2001). Third base. *American Scientist,* 89, 490–494.

Helfrich, H. (1996). Psychology of time from a cross-cultural perspective. In H. Helfrich (Ed.), *Time and mind* (pp. 103–108). Seattle, WA: Hogrefe & Huber Publishers.

Hitch, G. J., & Towse, J. N. (1995). Working memory: What develops? In F. E. Weinert & W. Schneider (Eds.), *Memory performance and competencies: Issues in growth and development* (pp. 3–21). Mahwah, NJ: Erlbaum.

Hofman, M. A. (1988). Size and shape of the cerebral cortex in mammals II. The cortical volume. *Brain, Behavior, and Evolution, 32,* 17–26.

Kareev, Y. (2000). Seven (indeed, plus or minus two) and the detection of correlations. *Psychological Review, 107,* 397–402.

Kavanau, J. L. (2002). Dream contents and failing memories. *Archives Italiennes de Biologie, 140,* 109–127.

Kompass, R. (in press). Universal temporal structures in human information processing: A neural principle and psychophysical evidence. In C. Kaemback, E. Schröger, & H. Müller (Eds.), *Psychophysics beyond sensation: Laws and invariants of human cognition.* Mahwah, NJ: Erlbaum.

Krahe, R. Kreiman, G., Gabbiani, F., Koch, C., & Metzner, W. (2002). Stimulus encoding and feature extraction by multiple sensory neurons. *Journal of Neuroscience, 22,* 2374–2382.

Levine, R. V. (1996). Cultural differences in the pace of life. In H. Helfrich (Ed.), *Time and mind* (pp. 119–140). Seattle, WA: Hogrefe & Huber Publishers.

Libby, L. K., & Neisser, U. (2001). Structure and strategy in the associative false memory paradigm. *Memory, 9,* 145–163.

Lustig, C., & Hasher, L. (2002). Working memory span: The effect of prior learning. *American Journal of Psychology, 115,* 89–101.

Luck, S. J., & Vogel, E. K. (1997). The capacity of visual working memory for features and conjunctions. *Nature, 390,* 279–281.

Lyons, R. G. (2001). *Understanding digital signal processing.* Upper Saddle River, NJ: Prentice-Hall.

Merker, B. (in press). Cortex, countercurrent context, and the logistics of personal history. *Cortex.*

Michon, J. A. (1996). The representation of change. In H. Helfrich (Ed.), *Time and mind* (pp. 87–100). Seattle: Hogrefe & Huber Publishers.

Miller, G. A. (1956). The magical number seven, plus minus two: Some limits on our capacity for processing information. *Psychological Review, 63*, 81–97.

Miyake, A. (2001). Individual differences in working memory: Introduction to the special section. *Journal of Experimental Psychology: General*, *130,*163–168.

Mountcastle, V. B. (1997). The columnar organization of the neocortex. *Brain, 120,* 701–722.

Neath, I. (1998). *Human memory.* Pacific Grove, CA: Brooks/Cole.

Neisser, U., & Hyman, I., Eds. (2000). *Memory Observed: Remembering in Natural Contexts.* Second edition. New York: Worth Publishers.

Nora, P. (1996). General Introduction: Between memory and history. In P. Nora (Ed.), *Realms of memory: Rethinking the French past, vol. 1: Conflicts and Divisions*, translated by A. Goldhammer. New York: Columbia University Press.

O'Connor, R. C., & Glassman, R. B. (1993). Human performance with a seventeen-arm radial maze analog. *Brain Research Bulletin, 30,* 189–191.

Pierce, J. R. (1992). *The science of musical sound.* Revised edition. New York: W. H. Freeman and Company.

Plomp, R. (1976). *Aspects of tone sensation.* Academic Press..

Pollack, I. (1952). The information of elementary auditory displays. *Journal of the Acoustical Society of America, 24,* 745–749.

Richardson, J. T. E., Engle, R. W., Hasher, L., Logie, R., Stoltzfus, E. R., & Zacks, R. T. (1996). *Working memory and human cognition.* New York: Oxford University Press.

Richelle, M. (1996). The expanding scope of the psychology of time. In H. Helfrich (Ed.), *Time and mind* (pp. 3–20). Seattle, WA: Hogrefe & Huber Publishers.

Rosenzweig, M. R. (1996). Aspects of the search for neural mechanisms of memory. *Annual Review of Psychology, 47,* 1–32.

Rosenzweig, M. R., Breedlove, S. M., & Leiman, A. L. (2002). *Biological psychology: An introduction to behavioral, cognitive, and clinical neuroscience.* Third edition. Sunderland, MA: Sinauer.

Rossing, T. D. (1990). *The science of sound.* Second edition. Reading, MA: Addison-Wesley.

Saaty, T. L., & Kainen, P. C. (1986). *The four-color problem.* New York: Dover.

Schack, B., Vath, N., Petsche, H., Geissler, H.-G., & Möller, E. (2002). Phase-coupling of theta-gamma EEG rhythms during short-term memory processing. *International Journal of Psychophysiology, 44,* 143–163.

Shastri, L. (2002). Episodic memory and cortico-hippocampal interactions. *Trends in Cognitive Sciences, 6* (4), 162–168.
Shettleworth, S. J., & Krebs, J. R. (1986). Stored and encountered seeds: A comparison of two spatial memory tasks in marsh tits and chickadees. *Journal of Experimental Psychology: Animal Behavior Processes*, *12*, 248–257.
Shaw, G. L. (2000). *Keeping Mozart in mind.* San Diego: Academic Press.
Singer, W. (1993). Synchronization of cortical activity and its putative role in information processing and learning. *Annual Review of Physiology, 55*, 349–374.
Wallin, N. L., Merker, B., & Brown, S., Eds. (2000). *The origins of music.* Cambridge, MA: MIT Press.
Watts, D. J., & Strogatz, S. H. (1998). Collective dynamics of 'small-world' networks. *Nature*, 393: 440–442.
Watts, D. J. (1999). *Small Worlds*. Princeton, NJ: Princeton University Press.
Waugh, N. C., & Norman, D. A. (1965). Primary memory. *Psychological Review, 72*, 89–104.
Weinert, F. E., & Schneider, W. (Eds.) (1995). *Memory performances and competencies: Issues in growth and development.* Mahwah, NJ: Erlbaum.
Weiss, V. (1992). The relationship between short-term memory capacity and EEG power spectral density. *Biological Cybernetics, 68,* 165–172.
White, E. L. (1989). *Cortical Circuits*. Boston: Birkhäuser.
Yaglom, I. M., & Boltyanskii, V. G. (1961). *Convex figures*. trans. P. J. Kelly & L. F. Walton. New York: Holt, Rinehart & Winston.
Zangwill, O. L. (1972). Remembering revisited. *Quarterly Journal of Experimental Psychology, 24,* 123–138.

Chapter 11: Invariants in mental timing: From taxonomic relations to task-related modeling

HANS-GEORG GEISSLER and RAUL KOMPASS

Abstract

This chapter is primarily intended to be an introduction to the Taxonomic Quantum Model TQM (Geissler, 1987, 1992) for those interested in the temporal organization of perception and cognition but not yet familiar with the approach. Derived through comparative evaluation of critical time periods from various research domains and tasks, TQM maintains that mental processing is organized in finite ranges of discrete periods which in turn are multiples of an absolutely smallest quantal period $Q_0 \approx 4.5$ ms. We illustrate this by pinpointing common temporal structures in data from audition, motion and time perception, and reaction-time performance. In an analysis of most recent evidence in apparent motion, we demonstrate an extraordinary agreement among individuals in basic quantal periods and a precision of timing of up to $\pm$ 1 ms and beyond. Physiologically, this precision can only be compared to that of inter-spike intervals in the forebrain of behaving apes as observed by Abeles and co-workers (e.g. Abeles et al., 1993). Further findings, in particular in duration discrimination, suggest that quantal timing is characteristic of modes of functioning of a general-purpose timing system which is also operative under the more frequent circumstances in which no obvious signs of temporal quantization are observed. Thus formulations of super-ordinate rules such as TQM can be expected to provide tools for converging research strategies that help to bridge gaps between traditionally distinct areas in the time-mind domain.

"Magic numbers" in mental timing?

As a system of discrete temporal characteristics and hypothetical relations among them, the Taxonomic Quantum Model (TQM) deviates from most other approaches toward time-mind relations in that it departs from strategies closely resembling taxonomic procedures through which the life sciences and chemistry have made their historical careers. As a consequence, in its current state of development, TQM is not a predictive theory and, thus, does not compete with specific models developed for particular phenomena of mental timing. In the following, pain will be taken, however, to show that the TQM rules about numerical and structural invariants reflect elements of a common ground in which many different manifestations of time-mind relations root. In this perspective, it is but natural to expect that specific modeling will ultimately benefit from the knowledge about such invariants. Even more, their systematic exploration across domains and paradigms should prove to be a prerequisite of any forthcoming comprehensive theory of mental timing.

The best short cut to the approach is through the history of its emergence: The first impulse towards a taxonomic turn arose in the early 1970s from the discontent of the first author with the fact that in simple psychophysical tasks data could indistinguishably be fitted to models resorting to widely differing, often enough even contradicting, assumptions. In his research in visual recognition, to circumvent this difficulty, he introduced a technique of "chronometric cross-task comparison." The main idea was to disambiguate models by temporal parametrization, thereby postulating invariance of time parameters under variation of stimulus parameters and task constraints (see e.g. Geissler et al., 1978, Geissler & Puffe, 1982; Geissler, in press, for a review). At that time another research group at the same institute did something similar by fitting latencies in standardized reasoning tasks to predicted numbers of operations (e.g. Klix & van der Meer, 1978). The estimates from the two lines of studies yielded a surprising picture: There seemed to exist small "bands" of operation times centering at around 55, 110 and 220 ms, thus exhibiting near-doubling relations. As a datum from the literature which fitted into this regularity the asymptotic value of 36.5 ms determined by M.W. Kristofferson (1972) came to mind which up to the first decimal is 1/3 of 110 ms. Taken together, these four values suggested a system of "magic numbers" as depicted in Figure 1a. Herein a period of 110 ms represents something like a "prototype duration" from which the rest of periods derives by either integer division or multiplication.

It took not too long to realize that this scheme was practically identical with the one of Figure 1a* advanced many years before by Stroud (1955), who adopted 111 ms for the ticks of a central clock. Identical—but with an important reservation: Stroud's data base, consisting of time thresholds and integration times taken from the literature, was quite different from ours. This suggested significance of the "magic numbers" beyond performance in recognition tasks.

A next step toward taxonomic reasoning resulted from an, at first glance, disconcerting observation: In the most important example, a cascade of discontinuities as demonstrated by von Békésy (1936) in the absolute thresholds for low-frequency sound, Stroud's rule was found to account for less than 50 per cent of the distinct breaks evident in the data. Somewhat later, it became obvious that all data accorded also with another ordering principle, namely that the critical period durations can be represented as integer multiples of some elementary unit. In visual recognition, evidence of such a multiplicative law emerged from a reanalysis of data from Vanagas et al. (1976) obtained in experiments with eight subjects. Employing sets of patterns derived from a prototype pattern through deletion of segmental lines, these authors had shown that percentage-correct rates after training accorded with staircase functions of presentation times exhibiting steps of approximately equal lengths. From various fit procedures for step lengths, Buffart and Geissler came up with an largest common denominator (*l.c.d.*) of 9.13 ms (see Geissler, 1985a) showing a standard deviation of 0.86 ms across individuals. It turned out that the four above-mentioned periods, although partly many times larger than this small period, can be represented as integer multiples of it, with nearly absolute precision: $4 \times 9.13 = 36.5$; $6 \times 9.13 = 54.8$; $12 \times 9.13 = 109.6$; $24 \times 9.13 = 219.1$.

Of course, this might have been some strange coincidence. Yet, later, chronometric analyses seemed to support a modular unit of some 9 ms (see Geissler, 1985a; Puffe, 1990; Bredenkamp, 1993). In Figure 1b, the scheme of Figure 1a for recognition data is supplemented by this elementary period and by a triplet of longer periods from a more recent item-recognition study with complex verbal material by Petzold and Edeler (1995).

A 9-ms fine-graining of time periods *per se* would rarely lead to any interesting modification in specific modeling. It becomes important only in conjunction with a property which Stroud's fractionation rule tried to acount for, namely that from the potential set of multiples only a small regular subset is observed. From Figure 1b one might extrapolate the selection rule that up to a factor of 5 all integer multiples of 9.13 ms actually occur and that beyond this point the quantal unit is enlarged to six times that value. In other words, the figure suggests a form of package or "chunk" formations to hold for quantal time periods.

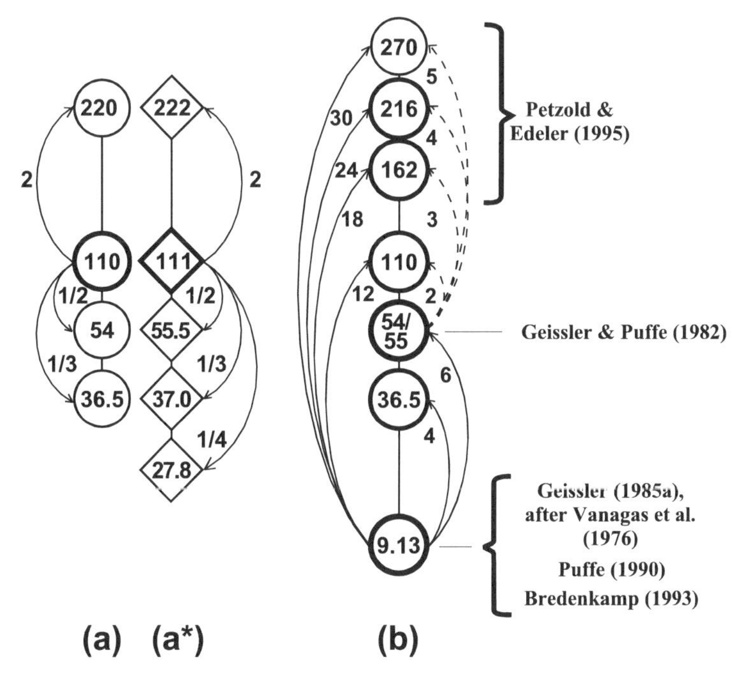

Figure 1: Critical periods (in ms) represented by multiplication and fractionation of a prototype (a, a*), and multiplication of an elementary epoch (b).

Surely, a useful taxonomy of critical time periods requires a much broader basis than a couple of estimates from early recognition experiments, the more so as the corresponding raw data seem to be lost. In the following, we shall, therefore, first provide a summary of the principal argument which has led to the formulation of more general rules. In a second step, we will outline novel evidence obtained in experiments designed under the guidance of this formulation and discuss consequences thereof that seem to enforce a revision of tenets on mental timing that have been taken for granted over decades of research.

Toward a proper set of rules

Invariances across paradigms and domains: Common principles of temporal organization

A major drawback of schemes based on paradigms in visual recognition like that of Figure 1b is that the validity of the values depends on the correctness of the models employed for estimation in each particular task. Fortunately, the situation becomes much simplified when referring to evidence on quantal patterns from other domains. A good illustration of this is provided by the early experiment of von Békésy (1936) on low-frequency sound to which Stroud (1955) had referred in suggesting his rule. Using a descending method of limits, Békésy' found a cascade of sharp breaks in the absolute threshold as a function of frequency. Conversion of the break frequencies into period durations yields the following 11 period durations (in ms): *222 (2.9)*, 166.7, 133.3, *111.1 (1.5)*, 90.9, 71.4, *55.6 (0.8)*, 45.5, *35.7 (0.8)*, 31.3, and 26.3. Here italics mark values that closely agree with four of the five recognition-based estimates within the range of overlap, the numbers in brackets denoting the absolute deviations from the latter.

To see what this reveals, we have to note that, in contrast to the recognition-based periods, all values of von Békésy's series have been determined within one single experiment without referring to any specific model assumptions, also, that the precision of most of them (with the possible exception of the three longest periods, because of 0.5-Hz rounding) is on the 1-ms level. Because of these properties, the close congruence of four values corroborates to some even if not readily quantifiable degree the factual relevance of the regularity found in the recognition-based estimates.

There are also discrepancies to note. Von Békésy's break series itself contradicts the asserted rule: A period of 31.3 ms clearly violates the assumed 9-ms module, and the close spacing of period durations >55 ms is at variance with the above chunking assumption. Yet the two data sets together justify a modified assumption about quantal graining: Regression yields the largest common denominator (*l.c.d.*) 4.6 ms, which is nearly exactly one half of 9.13 ms. Note that, in terms of hypothetical quanta, a period of such duration represents the next smaller candidate of a "true" elementary "time quantum" which is compatible with the recognition data. In the following, let us adopt provisionally the ("ideal") value of Q_0 = 4.565 ms for this *time quantum hypothesis*.

The data of von Békésy's early experiment, obtained from one single subject, do not permit further exploration of this hypothesis, nor do they provide a sufficient basis for deriving a proper rule for the selective occurrence of periods. Both goals require extraction of invariants from a larger sample of data sets cutting across domains and paradigms traditionally treated quite apart from each other. Table 1 may serve to explain an important step in this direction.

1	2	3	4	5	6	7
Békésy (1936)	Geissler et al. (1999, 2002)	Treisman et al. (1990)	Dehaene (1993)	Recognition study	Predicted (ideal)	Integer Multiple
		286 ± 10 (2)		*267*	273.9	30 × 2
222 ± 6 (4)		**222± 6** (2)		*216*[+]	**219.1**	**24 × 2**
		182 ± 4 (2)			182.6	20 × 2
166.7±3.5 (4)				*163*	164.3	18 × 2
133.3±2.2 (7)					132.4	29*
111.1 (12)	**108**(38);**110**	**111.1** (4)		*110*	**109.6**	**24**
90.9 (6)		87.0 (4)			91.3/86.7	20/19*
71.4 (5)		74.1 (3)			73.0	16
		69.0 (2)			68.5	15
		62.5 (2)			64	14
55.6 (16)	**54**, (16); **55**	**55.6** (2)		*54*	**54.8**	**12**
45.5 (7)	45 (8)				45.6	10
35.7 (9)	**37** (4)		**36.6 ± 1.3**(3)	*36.5*	**36.5**	**8**
31.3 (2)					32.0	7
26.3 (9)	**27**, (9); **27**		**28.4 ± 0.8**(2)		**27.4**	**6**
	22 (12)		21.3 (1)		22.8	5
	18 (2)				18.3	4
			13.5; 15,1 (2,1)		13.7	3
	9 (52)			9.13	9.1	2
	5 (7)				4.5	1

Table 1: Critical time periods (in ms) from four distinct paradigms as compared to multiples of a quantal unit of 4.565 ms. [+]Mean of three estimates. Further explanation in the text. *The factors 29 and 19 are prime numbers and, thus, violate "double determination" (see below). Note, however, that they are positioned in the middle between pairs of closely packed admissible values 30/28, and 18/20, respectively. This points to the possibility that periodic stimulation may have caused simultaneous resonance of the corresponding physiological carriers. Whether this assertion is sound can be decided only by careful experimentation with more densely spaced periodic or intermittent "single-shot" stimuli (see next section).

Besides the von-Békésy (1936) data in column 1, for the purpose of illustration, it includes in column 2 data from methodologically similar recent experiments in apparent motion by Geissler et al. (1999) and Jost et al. (2002), representing ISIs which are critical for the transition from perceived motion to flickering. Both experiments are reviewed in some detail further below. These results are contrasted by data from experiments that differ strikingly from the former in the methodologies employed: The data from Treisman, Faulkner, Naish, & Brogan (1990; see also Treisman, Cook, Naish, & MacCrone, 1994), in column 3, bring the domain of time perception into focus. Those by Dehaene (1993), in column 4, bridge the gap from complex recognition to more elementary decision tasks. All durations were recalculated from critical frequencies determined in these studies. Column 5 and columns 6 and 7 are reserved for comparison with the recognition-based data and for a check of the time quantum assumption based on them.

Briefly, the periods reported in Treisman et al. (1990; see also 1994) from an experiment with six subjects were inferred from the impact of periodic clicks on duration judgments. In terms of frequencies, the data represent maxima of underestimation, so-called dips, here conceived as substitutes of the theoretically significant neutral points which cannot be uniquely identified. The margins in the table represent maximal errors > 1 ms that may have resulted from the discrete spacing of click frequencies in 0.5-Hz steps.

The estimates from five subjects in Dehaene (1993) are based on surface periodicities in reaction time (RT) distributions in four two-choice discrimination situations, including simple visual or auditory stimuli, and conjunctions of those. Improving a rationale of Pöppel (1970), the estimates were obtained by performing fast Fourier transforms (FFT) and testing statistical significance of the resulting components in a range from 20 to 100 Hz. In Table 1, to reduce a possible blurring due to limited Nyquist resolution, from the 5 (subjects) × 4 (tasks) = 20 estimates only 9 periods of Bonferroni-corrected significances $p < 0.01$ and $p < .002$ have been accepted for inclusion. Error margins are given in two cases in which maximum possible errors by limited resolution may have exceeded ±1 ms.

The data from the four experiments are complemented by numbers in brackets denoting additional indicators to be interpreted as measures of prominence. In the order of columns these are: Height of break; height of mode in excess of 30 considered as noise level; depth of dip (all in arbitrary units), and number of cases.

Inspection of the first five columns of Table 1 reveals a clear cross-paradigmatic regularity marked by bold letters: Despite dramatic differences in the range extensions of the individual data sets, all data converge in five critical clusters in the vicinity of the values 220, 110, 55, 36/37, and 27 ms. Note that these epochs are exactly the ones of Stroud's scheme (Figure 1a*). A specific significance of these values is also indicated by the concentration of high prominence weight on these values in the hearing and motion experiments. These results strongly suggest invariant structures of temporal organization across distant paradigms employing indicators as different as operation time estimates, jumps in sensory threshold values, quantal bounds between alternating percepts, or driving effects in perceived duration.[19]

[19] We wish to emphasize that multiple comparison of quantal values is not the only way to pinpoint invariances. In special cases, agreement of critical periods across paradigms has already been demonstrated more directly. A brilliant example provide Burle and his colleagues (this volume, chapter 12) by showing nearly identical effects of periodic click stimuli on judged duration and on operation periods in item recognition.

To allow a check of the time quantum hypothesis, in columns 6 and 7 closest-matching multiples of Q_0 and the corresponding integer factors, respectively, are given, where the recognition-based value 9.13/2 = 4.565 ms is adopted as ideal estimate. Matching the 36 values ≤ 111 ms against this quantal lattice permits rejection of the chance hypothesis with $p < 0.001$ in one- and two-tailed testing. The time quantum assumption is thus corroborated without any additional fitting procedure. Because of a lack of precision, no analogous evaluation is possible for periods > 111 ms. However, the three values clustering in an interval of ± 2 ms around 220 ms justify a doubling hypothesis relative to the values of ~110 ms.

Translating the cross-paradigmatically invariant structures into multiples of Q_0, the findings suggest a preferred status of the following multiples: 48 (= 4 × 2), 24, 12 (= 24/2), 8 (= 24/3), and 6 (= 24/4). This agrees with Stroud's rule, with the additional constraint that admissible quantal periods are simultaneously integer multiples of a smallest quantal period. Notably, the periods observed in beta motion (factors 1, 2, 4) and RT periodicities (factor 3) conform with the remaining fractions 4, 3, 2, and 1 of the "generating multiple" 24.

Intuitively, this modified Stroud rule is a great step forward. However, in relation to the remaining data there are two serious drawbacks to note: First of all, the rule is even more restrictive than the fractionation part of Stroud's original proposal. As becomes obvious by closer inspection of the von Békésy (1936) and Geissler et al. (1999) data, in the range of durations < 110 ms, there are further, likewise regular, sub-series to be accounted for. Second, 5 of the 8 periods of durations > 110 cannot be represented as integer multiples of 12 Q_0 as "reference" period. This concerns even two of the Petzold-and-Edeler (1995) estimates which so beautifully fitted into the chunking scheme of Figure 1b.

The quantal-range constraint and pertaining empirical relations

The solution TQM offers to these seeming contradictions (see Geissler, 1987; 1992; also 1985b) can be considered as a generalization or at least an analogue of the psychophysical principle of relative-range constancy.[20] According to Teghtsoonian (1971), this principle expresses itself in the fact that for all sensory continua, in terms of output magnitudes, the ratio of the largest to the smallest quantity is a constant of around 30. About the same value is obtained for the ratio of a magnitude to the corresponding *jnd*, the so-called Subjective Weber Law.

The generalization of the principle in the realm of quantal timing is the *quantal-range constraint*. To see how this analogue reads, consider first the assumed smallest period Q_0. For integer multiples n × Q_0, consistency with the relative range constraint implies n ≤ M, with M being a constant of the hypothetical value 30. It follows that periods of durations in excess of 30 × Q_0 ≈ 137 ms cannot be represented within this smallest possible range. To account for such periods, we have to assume larger ranges with correspondingly larger admissible smallest quantal periods to be opera-

[20] The exact relation and even the validity of the principle are of no essential importance here. In the present context, the focus is solely on a simple introduction of the quantal-range concept.

tive. To retain consistency with the time quantum assumption, these periods must be integer multiples of Q_0 or, formally, $Q_q = q \times Q_0$ with integer q must hold. Thus, in general, the maximum extension of any quantal of periods T_i belonging to it is given by $q \times Q_0 \leq T_i \leq M \times q \times Q_0$. Note that the lower bound $q \times Q_0$ also defines the smallest possible distance between admissible periods within a range. For this reason we will speak of it as the *quantal resolution* within a given range.

Of course, in the actual development, this abstract definition resulted from a variety of empirical relationships suggesting a range ordering of quantal periods with upper bounds maximally at 30 times the value of quantal resolution. Of specific relevance for recognition data was an empirical relation discovered by Cavanagh (1972). He found that operation times in item recognition plotted against reciprocal memory spans (as measures of item complexities) can be approximated by a straight line. Regression from the conglomerate of data from various sources yielded a slope of 243 ms to be interpreted as the average time necessary for processing a full short-term memory. In relation to the quantal values of Table 1, this demonstrates a difference in the status of data subsets. While the periods of Dehaene (1993) and periods of durations < 110 ms from column 5 correspond by their nature to the primary data of the Cavanagh relation, i.e. to operation times, those of the Petzold and Edeler (1995) study are of the character of full memory cycles of which the slope of the line represents a compound. Indicative of an upper bound at M = 30 is a corrected estimate of 274.4 ms (Petzold et al., 1999) for the slowest subgroup which is in excellent agreement with the predicted maximum value of $30 \times 9.13 = 273.9$ ms.

Even more convincing support of quantal range structuring came from studies in duration discrimination by A.B. Kristofferson and co-workers. Figure 2 depicts the results of a single-case study (Kristofferson, 1980). In the experiment, the subject was to judge whether test intervals represented by pairs of clicks appeared long or short relative to the range of presented intervals. In the diagram, discriminability is plotted in quanta of Kristofferson's real-time criterion model against base duration. The branches unfolding in the course of extensive training from the original Weber line display a striking range structure of steps increasing in width and height which accords to a doubling law with cuts at base durations of about 24 times the corresponding quantal resolutions. If Kristofferson's model can be taken for granted, the ordinate intercepts should estimate corresponding asymptotic levels of quantal resolution. Regression yields the values 9.5, 19, 38 and 76 ms, which generally are only about 4 per cent above the predictions of 9.1, 18.3, 36.5 and 73.0 from rule $2^n \times (2Q_0)$.

The example again indicates that ranges may terminate at multiples below the theoretical maximum of 30, apparently with a preference of a factor of 24. This agrees with the later findings in apparent motion: In beta motion, an *l.c.d.* of 4.5 ms corresponds to a terminating mode at around 110 ms (Geissler et al., 1999). Similarly, in unpublished data from gamma motion, the terminating mode is found at 220 ms cISI, and the corresponding *l.c.d.* is 9 ms (see Kompass & Geissler, 2001). Note, however, that in these cases the ranges are fixed by innate constraints of the perceptual apparatus. By contrast, quantization in duration discrimination emerges as a result of training. Base durations of up to 1.5 seconds in the present example indicate that as a result of extensive practice quantal range structures may extend over much larger intervals. That under certain conditions also the theoretical maximum of relative extension can be attained is known from a more recent study (Kristofferson, 1990): In the so-called pulse-train paradigm, subjects were to judge a deviant, following a series of isochronal pulses. For two subjects, plateaus of constant discriminability were obtained terminating at 278 and 286 ms, repectively. The respective quantum sizes were 8.6 and 10.3 ms, averaging 9.45 ms. With the measured range extensions this yields 282/9.45 = 29.8, which is close to 30.

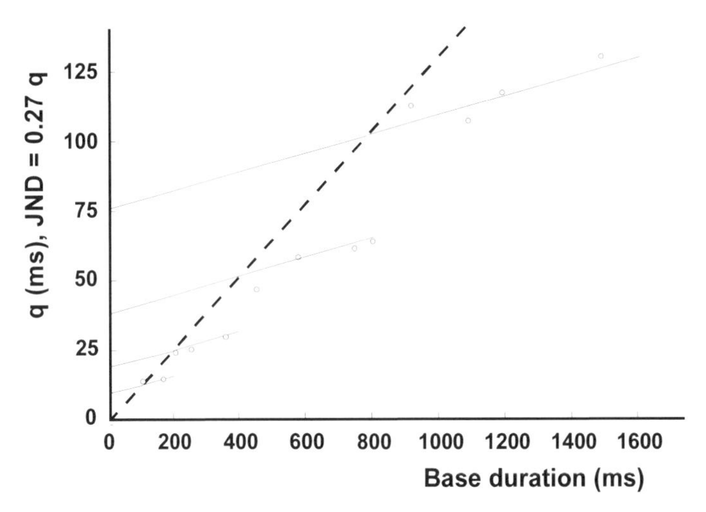

Figure 2: Discriminability as a function of base-time: During training step-like deviations unfold from the original Weber's Line (dashed diagonal).

The TQM rule of "double-determination"

These findings bring us to the core assumptions of TQM: Suppose the quantal-range hypothesis is sound. Then we have not to worry any further on how to account for periods longer than a suggested generator period. Apparently, the only option avoiding arbitrary additions reads that in every concrete case the upper range bound itself represents the "generative" period of which all other periods are integer-ratio segments. From the evidence discussed follows also that, in order to explain the real manifold of quantal periods, more than one generator is to be assumed. As one easily realizes, the set of these generators is completely determined by what we call "double determination," i.e. the fractionation rule in conjunction with the condition that admissible periods are integer multiples of the quantal unit operative in a given case with $M = 30$ as upper bound. Specifically, the integer factors corresponding to a generator period must be elements of the set $16 \leq N_i \leq 30$ of multiples, because no element of this set is divisor of another, while all other multiples $n_i < 16$ are divisors of one (or more) of those. From the 15 candidate factors, the primes17, 19, 23, and

29, not allowing uniform segmentation, can be dropped. The set of relevant periods can further be narrowed down by the observation that 22 and 26 each have only one large divisor. The remaining 9 elements are still far from being of equal significance. Figure 3 may help to illustrate this. In the picture, the solid vertical lines mark the lower bounds of the total of four sub-ranges of the property that no elements are divisors of the remaining ones.

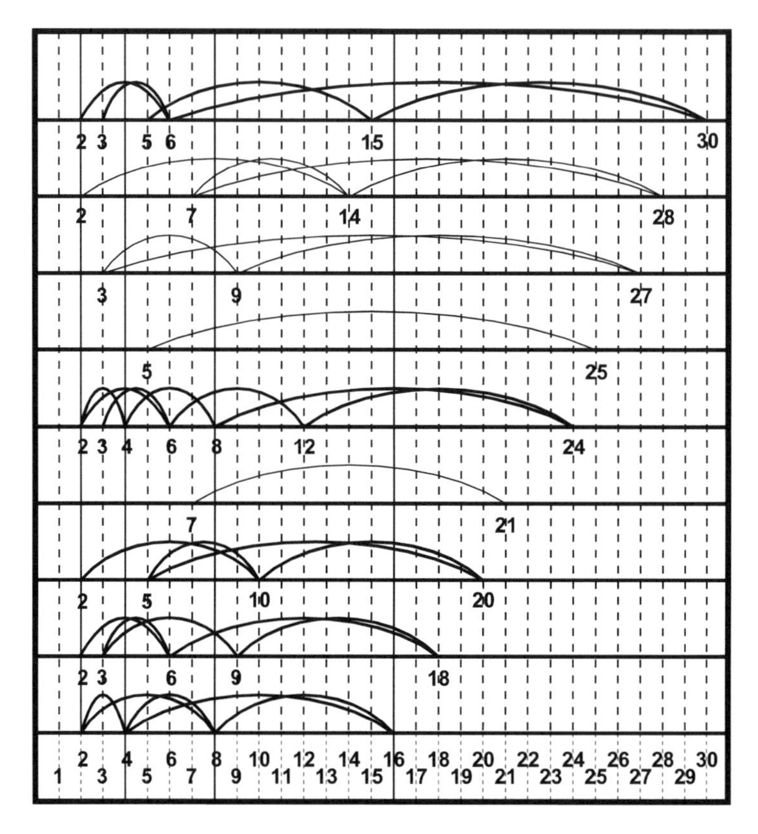

Figure 3: Illustration of "double-determination."

Suppose the rule of "double determination" reflects to a good approximation structures resulting from the action of inner constraints of brain functioning. Then the factors determining the autonomous preference order among admissible periods should also be inherent to it. The simplest characterization of the subset of generators which could account for observed inequalities is by the number of divisors which is the same as the number of supported segmental periods. With 6 integer divisors each, 24 and 30 are on top of the list, followed by 20 and 28 with 4 divisors. 16 and 18, with 3 divisors each, occupy a medium position while the rest of potential generators have only two or one divisor. Even this primitive characterization accords well with findings from Table 1: with the prevalence of the factor 24 and its divisors, the patterning of dominance indicators, and with the preferred occurrence of the multiples 16, 24 and 30 of $2Q_0$ in complex item-recognition. Very probably, however, it is not the number of possible segmentations *per se* that matters, but the power to flexibly combine various different segmentations in processing. This suggests emphasis on hierarchical relations, in Figure 3 indicated by arcs.[21] Independent of the particular criterion one may choose, inclusion of hierarchical relations leads to a much sharper prevalence of a few generators: In particular, 24 dominates all the other multiples. As indicated by bold arcs in Figure 3, only four of the possible generators, namely 30, 24, 20, 18 and 16, show an appreciable degree of interconnectedness with smaller multiples. These generators are, thus, expected to prevail in observed performance.

A provisional oscillator interpretation

In the great majority of cases, mental timing in general and quantal effects in particular have been interpreted in terms of brain oscillations that serve the function of internal clocks. At this point, a brief discussion seems in order on how the postulates of double determination just outlined fit into a framework of oscillatory mechanisms and which architectural specificities such a representation supports. A corresponding interpretation (Geissler, 1991, 1997; cf. Geissler & Kompass, 1999) was encouraged by a striking correspondence between the hypothesized sub-range of generators and the band width of the most pronounced oscillations in the brain: With the boundaries N = 16 and N = 30, for the finest possible resolution of 4.57 ms the assumed generators span a range of period durations from 73.1 to 137.1 ms, or, converted into frequencies, from 7.3 to 13.7 Hz. Notably, although these predictions derive from a purely behavioral base, they agree with the empirical definition of the alpha band.

The agreement motivated the hypothesis that oscillatory activity in this range plays a fundamental role in the generation of faster, functionally significant oscillations in other bands (Geissler, 1990; 1997). We note *in passing* that this does not imply that the well-known high-amplitude oscillations showing up just in stages of functional relaxation are directly involved. Rather, the involved portions of periodic excitation may make up only a very small, entrained and specifically distributed portion branching from the total activity within the alpha band. An example relevant to timing was provided in an EEG study by Maltseva et al. (2000), employing a mental signal-anticipation task. For related independent developments, the reader is referred to work by Jensen (1998; cf. Jensen et al., 1998) building on a proposal by Lisman & Idiart (1995), and to a bulk of papers signaling a renewed interest in the functional significance of activity in the alpha band (see Basar et al., 1997).

Of particular significance for the purpose of the present discussion is an interpretation in terms of coupled oscillations which accounts quite naturally for the inner-range regularities of double determination. In this interpretation, group oscillations in different frequency bands are assumed to become bound together by phase-locking,

[21] For the sake of simplicity, for each multiple maximally the two largest divisors have been included.

in this way giving rise to hierarchies of critical periods. Figure 4 illustrates the principle by way of example. Because of their composition of components of slightly differing frequencies, group oscillations exhibit temporally limited coherence. This spontaneous relaxation captures the critical property of upper-range limitation. In conjunction with grouping by phase-locking follows a hierarchical organization of activity in *time-frequency windows*—in Figure 4 indicated by solid, dashed and dotted rectangular frames—with cutoffs at the points where coherence of the respective fastest oscillations expires. This automatically yields the already known regularities of within-range organization. Note, however, that beyond this the model exhibits properties, not yet explored in any greater depth, by which it may provide a starting point for further developments. An example are properties by which the model constrains the dynamics of between-range transitions: If a period to be represented exceeds the size of the largest period included in a current time-frequency window, enlargement of window size is mandatory. By contrast, if maximum period durations fall below a critical boundary, revision of window size is optional and may be delayed until a noticeable lack of discriminability enforces revision. Yet in both cases the exact structure of the established hierarchy should depend on the preceding one, because new oscillations that become included into a time-frequency window have to be compatible with still ongoing oscillations.

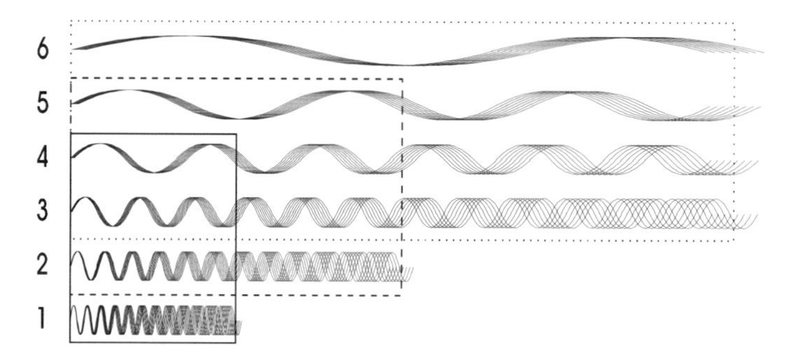

Figure 4: Illustration of a provisional oscillator model: Component group oscillations are assumed to form hierarchies through phase locking. Limited temporal coherence causes organization in time-frequency windows (solid, dashed and dotted frames).

In comparison to current clock models of mental timing there are three specificities to note: (1) In contrast to approaches adopting one central clocking device, sets of coactive oscillations are assumed. (2) Because of spontaneous relaxation, these oscillations resemble stop watches rather than continuously running clocks. (3) Different from single-clock models which necessitate additional special-purpose devices to account for situational specificity, it is maintained that (within the limits of domain and task specific constraints) the timing regime in any given situation emerges according to general rules of flexible organization.

Facets of double determination: Steps towards a refined analysis

In the remainder of this chapter, we provide evidence for our claim that—in deviation from familiar model-based strategies—a taxonomic structure like TQM may lead to novel facts and unexpected conceptual progress. To this purpose we review two experiments in apparent motion in some more detail. Although data from the first of these experiments were for convenience of illustration included into Table 1, motivating the rule of double determination, in actual fact it was carried out more than ten years after publication of the first versions of TQM (Geissler, 1985b, 1987), and thus represents an example of its success.

Quantal periods in beta motion: A method-of-limits experiment

In both experiments (long-range) beta motion was induced by periodically alternating stimuli, a condition which seems to be required for stable temporal quantization within narrow intervals of variation. As the inset of Figure 5 illustrates, the stimulus display consisted generally of two circular flashes of light presented in periodic alternation at two sides for an exposure duration ED. After each presentation, the field remained dark for some ISI. For appropriate combinations of ED and ISI, symbolically represented by the hatched area in the phase diagram below, a clear periodic forth-and-back motion of a patch of light is seen. The lower demarcating line is traditionally defined as "simultaneity threshold. " "Threshold", in this conceptualization, refers to the critical ISIs (cISIs) at which the motion percept is most likely to break down, giving rise to flickering-circle percepts at both sites.

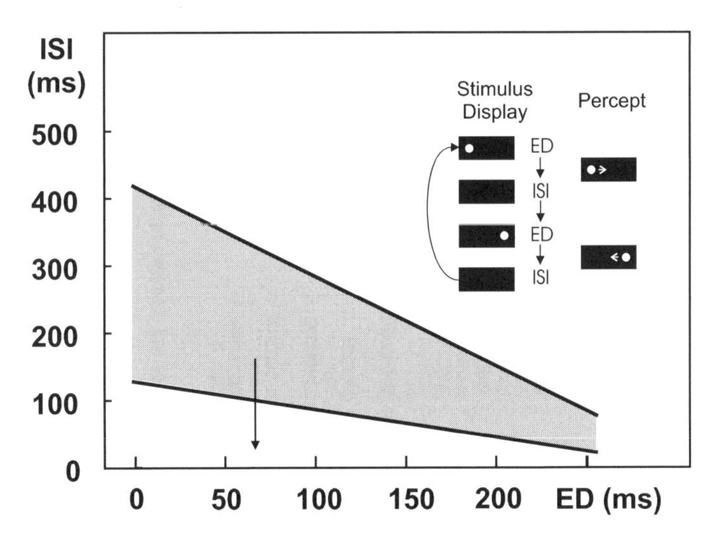

Figure 5: Hatched area: Conditions of stable beta motion. Inset figure: Scheme of stimulus presentation and the corresponding percept.

In the experiment of Geissler et al. (1999) (here Experiment 1), cISIs were determined with a method of limits by downward adjustment of ISI, in Figure 5 indicated by an arrow. There were two trials for each of 12 ED values and 3 spatial distances. cISIs were recorded in ms steps. 46 subjects participated in the experiment.

To explore mode positions, cISIs were collapsed across spatial distances and EDs. The resulting distribution is shown in the upper panel (a) of Figure 6. To check upon statistical significance of modes, various procedures were tried out with essentially identical results. The two horizontal lines indicate significance levels based on simulations.

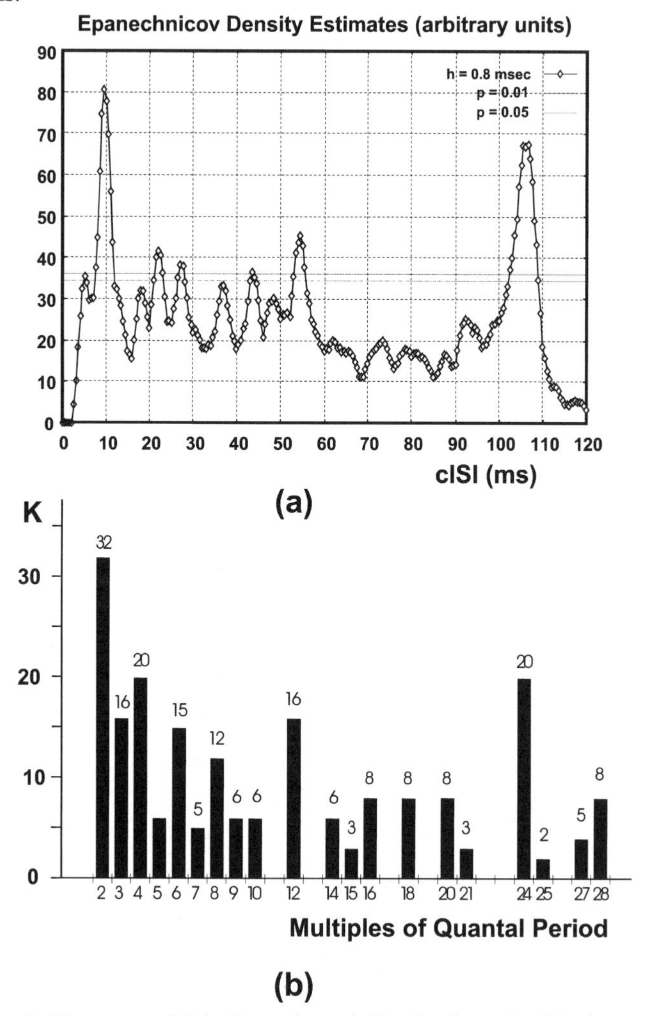

Figure 6: Upper panel (a): Smoothened distribution of critical ISIs (cISIs) at transition of seen motion to flickering. Lower panel (b): Theoretical frequencies.

Let us first apply the viewpoint of double determination: If including the non-significant—but correctly positioned—peaks at the multiples 4 and 8, and a flat maximum (in brackets) at the right flank of the peak at 9 ms—smoothed away in Figure 6a—, there are peak positions corresponding to the multiples 1, 2, (3), 4, 6, 8, 12 and 24 of Q_0 which are compatible with the full spectrum of multiples pertaining to 24 as hypothetical generator. Together with the multiples 2 and 4, the remaining modes corresponding to 5 and 10 form the full set of divisors of 20 as generator, while the weakly pronounced maximum at about 92 ms might—with some caution—be seen to represent the generating period itself.

The reverse statement, that all observed periods derive from the multiples 24 and 20 as generators, is, of course, not justified. Some of the modes could well belong to 18 or 30, with generators themselves, for whatever reasons, remaining hidden. Other generators could participate as well without any noticeable effects in excess of a certain "noise level." To surpass this fuzziness involved in statements about relationships to special generators, one would wish a theoretical prediction for the full distribution to be available as a basis of comparison. As suggested in Geissler et al. (1999), the oscillation interpretation considered above provides a starting point for such an approach. In Experiment 1, the full range of potential cISIs was traversed about evenly. Thus it appeared reasonable to suppose that the marked differences in the frequencies of breakdown are primarily caused by inner factors of temporal organization. Specifically, it was assumed that the relative frequency of motion breakdowns at a particular cISI corresponds to the cumulative frequency of that ISI value being contained in the set of possible (complete) hierarchies whose realization across individuals is supposed to be random. With the semi-empirical constraint of a maximum at N = 28 suggested by the upper truncation of the empirical distribution, this yields the profile depicted in Figure 6b. Considering the roughness of the approximation, the agreement with the empirical course, including the slightly asymmetric trend, is encouraging. It is mainly the "lines" corresponding to the assumed generator N = 20 that are empirically more pronounced. No plausible basis for a comparison is given for cISIs of durations close to Q_0.

Applications of the oscillation model can be considered as straightforward derivations from the regularities of double determination. Beyond this, the data obtained from a large group of subjects permit answers to fundamental questions left open in establishing the principle. Specifically, Q_0 was hypothesized on the basis of largest common denominators. A deeper understanding of the observed quantal graining would, however, require answers to the questions whether there is a genuine entity corresponding to Q_0 in mental processing, and whether temporal quantization is universal or subject to individual variation.

A fairly convincing answer to the first question came from an analysis of the individual cISI distributions demonstrating a periodic 4.5-ms modulation in the fluctuations between first trials and repetitions (Geissler et al., 1999). Figure 7 shows the same on the basis of a more recent evaluation of cISI differences for neighboring EDs.

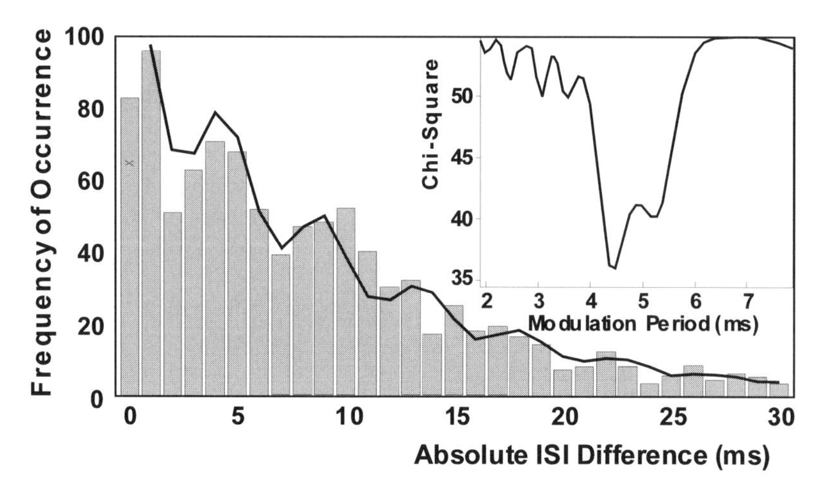

Figure 7: The distribution of differences between cISIs for neighboring EDs show a modulation with a period of ≈ 4.5 ms suggesting fluctuations in steps of this duration. Inset figure: Chi-square fit as a function of period duration.

The second question is readily answered: For tall, narrow modes to appear as observed in the full distribution of critical ISIs, most of the subjects must have agreed in the times of break-down with a precision of ± 1 ms. Although this statement draws upon surface properties of the cISI distribution, it is by no means trivial. A particularly surprising consequence follows when relating it to the stimulus regime: In Experiment 1, due to the cyclic alternation of ED and ISI quantal timing is *not* bound to evenly-periodic stimulus presentation as in the experiments of Békésy (1936) and Treisman et al. (1990). Moreover, since ISI was continuously adjusted, stimulus presentation not even accorded to a fixed rhythmic alternation. Together with the precision argument, this implies that the breakdowns of seen motion must be timed *anew* with a precision of ± 1 ms *in every single event* of cycling stimulus presentation.

Note that both these findings, strong agreement among individuals in the basic periods as well as precise *single-shot timing*, are not easily reconciled with currently prevailing ideas and knowledge about quantal timing. More generally, there is good reason to believe that they mark the necessity of a fundamental turn in the thinking about mental timing.

Challenging "pure oscillation" views: A digression

Before looking somewhat deeper into matters of task-specific implementation, let us face for a moment possible consequences to our understanding of the underlying neural processes. Quite apparently, any explanation of quantal periods in terms of oscillations encounters difficulty to account for a precision far beyond the limits of the smallest quantal period to be assumed. Also, the empirical variation of characteristic frequencies between individuals does not seem to support precise congruence of fundamental periods among individuals as suggested from behavioral data. Even when precisely timed carrier oscillations, making up only a small transient fraction of the total oscillatory activity, in the brain are assumed, it remains obscure how sufficiently precise on- and off-set timing could be accomplished. Obviously, this objection includes the above oscillator interpretation and its phase-locking principle.

How can this puzzle be resolved? What we propose in order to reconcile the seeming discrepancies is that double determination be conceived as an result of the interaction of two mechanisms. More specifically, we assume oscillation-like representations of time to interact with delay-like representations (Geissler, 1994; Kompass, 2002).

The existence of delays in the brain follows from plain knowledge about neural transmission times and refractority periods. That neural delays may play a fundamental role in the precisely timed coordination of brain processes was first suggested by the discovery of Abeles and co-workers (e.g. 1993, 1994), who in the forebrain of apes found inter-spike intervals of the order of some tenths to 500 milliseconds and more, reoccurring with a precision of ± 1 ms. According to Abeles, so-called synfire chains can by concatenation of elementary delays produce larger delay periods which thus should be automatically quantized. Also, the resulting delay periods can be elements for chunking into higher-order delays. Unlike oscillation-type representations, this does not require any form of coupling between independent constituents like oscillations of different frequencies, because by chunking into chains, smaller chains become and then are constituents of longer chains. Furthermore, triggering of delays can be accomplished with utmost precision at any time and can thus account for single-shot timing. Most interestingly, however, if assuming mechanism permitting stabilization of oscillatory activity at zero phase lag to synfire waves this cooperation yields directly the constraint that defines double determination. For details, also concerning the functional significance of recurrent synfire chains and further physiological evidence, the reader is referred to Kompass (2002).

Beta motion again: Towards the functional denotation of quantal modes

Besides these conclusions about possible mechanisms, the outcome of Experiment 1 has also led us to a rethinking of the *functional role* of quantal timing to be briefly outlined here.

Fingerprints of ultra-precise quantal timing as found in the experiment surface only under very specific conditions, a crucial requirement being stimulus presentation in continuous ED-ISI sequences. However, even when this basic requirement is met, whether tall modes are observed or not may depend on seemingly subtle changes of conditions. Among the factors whose inappropriate choice may flatten or wash out tall modes in the ISI distributions are start values, speed of adjustment and trial-to-trial context. On a more general level, this poses the question of whether quantal precision of perceptual change found in apparent motion represents only a certain functionally insignificant resonance effect induced by artificial conditions or whether it reflects a truly fundamental principle of functioning operative also in the majority of paradigms without obvious signs of temporal quantization. An answer to this question necessitates reference to a theory, however fragmentary it may be. Figure 8 displays the essence of an early suggestion of D.M. MacKay (1963) on perceptual updating which we believe can be generalized to complex percepts (see also Geissler, 1983) and accounts for temporal quantization.

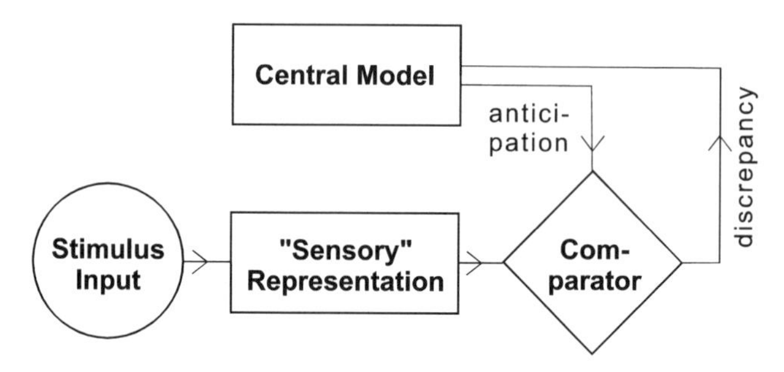

Figure 8: Scheme of perceptual updating after D.M. MacKay (1963): Percepts are considered as central models of reality which are matched against sensory evidence. Percept revision occurs only when the detected difference exceeds a discrepancy threshold. See text for further explanation.

In the vein of this scheme, we suppose that updating is initiated by discrepancies resulting from a comparison of predictions derived from an inner model of the distal stimulus with current sensory evidence (see Kompass & Elliott, 2001; cf. also Vanagas, 2001). Here the inner model is assumed to include the momentary percept itself as well as representations of its permanent framing through categorical abstractions such as prototypes and structural schemes. In brief, an application to timing in apparent motion reads as follows: Suppose that internal representations of time intervals refer—well in analogy to figural singularities (see Goldmeier, 1982)—to certain preferred quantal periods or standards. In instantaneous updating, there should then two major aspects be involved: First, the central model anticipates the current ISI with a precision that increases with decreasing distance of the modeled period to the standard which is active on a given occasion and which is co-determined by the modeling of the temporal stimulus structure. Second, updating occurs when discrepancies exceed a certain tolerance relative to the precision of this coding mechanism. It may either imply adjustment of the parameters of the central model without changing its structure or include qualitative revision. In apparent motion, this latter case corresponds to breakdowns which thus are expected to occur in moments of updating while the probability of breakdown may have nothing to do with the updating cycle as such. From the assumed modeling of temporal regularities it is understandable why maximum precision of breakdown at preferred values is attained only for regular stimulus sequences as used in Experiment 1. Complex interactions are to be expected with speed of adjustment: If adjustment is sufficiently slow, breakdowns resulting from cumulative deviations may be delayed to non-quantal values below the current standard. Fast adjustment, on the other hand, may for an ISI sub-range of high transition probability lead to discrepancies causing a sudden breakdown. Quantal breakdowns in close vicinity of the point of lowest tolerance correspond to a narrow window of conditions between these extremes.

Mechanisms employing inner standards for temporal coding require processes of activation and deactivation of standards. Recent findings for increasing and decreasing sequences of EDs may, with some caution, be interpreted as evidence of such processes (Jost et al., 2002): While mean cISIs for decreasing variation of ED (Series A, corresponding to increasing mean cISIs) matched closely those obtained in Experiment 1, mean cISIs for increasing EDs (Series B, corresponding to decreasing mean cISIs) were up to 20 ms smaller. There is one procedural difference to explain this astonishing asymmetry: Namely, only in Series B during adjustment the cISIs of the preceding trial are with high probability bypassed again. Thus, if breakdowns are related to inner time standards, their lack at those cISI values in the subsequent trial suggests that standards remain inhibited for a while after breakdown.

The weakest conclusion from the above considerations reads that a lack of tall quantal modes in critical ISIs does not exclude that precise quantal inner standards are active. Analogously, in other realms such as discrimination performance or choice reaction, the occurrence of quantal phenomena may prove to be indicative of a general role of such standards. For the present example of beta motion we quote provisional evidence which indicates that the coding assumption may apply also in the classical conditions of pairwise stimulus presentation. Since under these conditions at any trial the initial state (motion versus flickering) is indeterminate, experimental frequencies approximate the average chances of being in the state of seen motion rather than those of transition from motion to flickering. Data from extensive measurements under these conditions by Kolers (1964) are shown on the left-hand side of Figure 9. Note that the maxima in the plot against SOA = ISI + ED (“formation time”) agree roughly with cISI maxima of quantal breakdown in the method-of-limits experiment. The right-hand side of Figure 9 displays evidence from simulations based on a descriptive model. In the model, the increasing part of the functions is represented by a (bounded) continuous autocatalytic growth while a quantal-range structure is imposed by instantaneous decay after a precise delay (“storage”) period. After an approximate adjustment of the growth parameter, overall agreement with the family of empirical functions depends sensitively on the choice of the delay. Best congruence was found for a delay of $T = 120$ ms, which agrees closely with the upper limit $N = 28$ as assumed in Figure 6b[22].

Beside this numerical agreement, there are long tails showing up in the simulation as well as in the empirical data which represent a secret sign of delay-based quantal decays combined with continuous autocatalytic growth and thus give additional support to the modeling idea.

---

[22] To obtain congruence, the additional assumption had to be made that a preceding threshold-like process prevents seen motion from happening up to SOAs of around 30 ms. This period of the character of a pure refractivity ("threshold") does, however, not limit evaluation of the dynamic parameters.

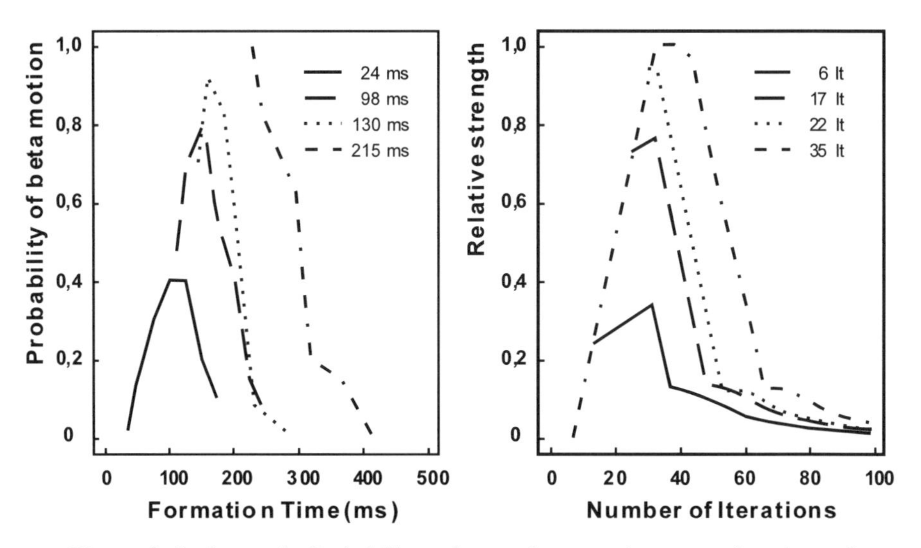

Figure 9: Left panel: Probability of seen beta motion as a function of stimulus onset asynchrony SOA ("formation time") with exposure time as parameter. Adapted from Kolers (1964). Right panel: Simulation assuming a constant delay period.

Consequences for task-specific implementation

From the above discussion it becomes apparent that the temporal properties at issue are not neutral against the cognitive content of a percept, as, for instance, an interpretation in terms of binding oscillations would imply (cf. Geissler & Kompass, 2001), but rather are part and parcel of it. Correspondingly, depending on the global temporal context of an experiment, precision of timing may strongly vary. When, as in the single-trial design, stimulus events do not repeat the same temporal pattern with utmost precision, perceptual updating will not occur with stable reference to one internal standard and, consequently, surface phenomena will show no obvious quantal graining. The assumed matching mechanism, thus, makes it conceivable why under most "normal" conditions surface phenomena exhibit no obvious quantal graining while still the same basic laws of quantal-range formation do apply.

In this view of temporal quantization, two features are specific to micro-genetic processes of perception: the automatic character of fast structure formation, and the matching criteria of the assumed updating mechanism which emphasize sameness and small deviations from it. Otherwise the conceptualization opens a door for generalizations which help to answer the question of how the same quantal invariants can be implemented in processes of higher-order cognition. A good example is provided by Kristofferson's (1980) findings in duration discrimination: Quantal steps in the discriminability function emerge as a result of extensive training under the specific constraints of stimulus presentation and task. At the same time, the lawful relation to

Weber's line testify to the existence of common invariants in quantal and non-quantal responding. Kristofferson's real-time criterion model can be seen as a useful approximation of intentional processing in the task. The role of a time standard is played by a representation of the stimulus ensemble established in the course of training. The assumed matching criterion is one of ordered discrimination. Within a quantal framework this signifies that the degree of deviation is not specified and, thus,—different from precise quantal values in perception—quantal levels of discriminability denote *intervals of uncertainty*.

To conclude, the condensed survey of this chapter has carried us from numerical list correspondences among empirical periods taken from distinct domains and paradigms to the view of a modular, general-purpose timing system operating in close interaction with specific content-related organizational processes. Suppose this view comes close to the truth, it should encourage common efforts of those engaged in research topics related to mental timing much to the benefit of their own specific fields.

References

Abeles, M., Prut, Y., Bergman, H., & Vaadia, E. (1994). Synchronization in neural transmission and its importance for information processing. In G. Buzsaki, R. Llinás, W. Singer, A. Berthoz, & Y. Christen (Eds.), *Temporal coding in the brain* (pp. 39–50). Berlin: Springer.

Abeles, M., Vaadia, E., Bergman, H., Prut, Y., Haalman, I., & Slovin, H. (1993). Dynamics of neuronal interactions in the frontal cortex of behaving monkeys. *Concepts in Neuroscience, 4*, 131–158.

Basar, E., Hari, R., Lopes da Silva, F. H., & Schürmann, M. (1997). Brain alpha activity and functional correlates. *International Journal of Psychophysiology, 26*, 1–482.

Bredenkamp, J. (1993). Die Verknüpfung verschiedener Invarianzhypothesen im Bereich der Gedächtnispsychologie. *Zeitschrift für Experimentelle und Angewandte Psychologie, 40*, 368–385.

Cavanagh, J. P. (1972). Relation between immediate memory span and the memory search rate. *Psychological Review, 79*, 525 – 530.

Dehaene, S. (1993). Temporal oscillations in human perception. *Psychological Science, 4*, 264–269.

Geissler, H.-G. (1983). The inferential basis of classification: From perceptual to memory code systems. Part 1: Theory. In H.-G. Geissler, H. F. Buffart, E. L. Leeuwenberg, & V. Sarris (Eds.), *Modern issues in perception* (pp. 87–105). Amsterdam: North-Holland.

Geissler, H.-G. (1985a). Zeitquantenhypothese zur Struktur ultraschneller Gedächtnisprozesse. *Zeitschrift für Psychologie, 193*, 347–362.

Geissler, H.-G. (1985b). Sources of seeming redundancy in temporally quantized information processing. In G. d'Ydewalle (Ed.), *Proceedings of the XXIII International Congress of Psychology of the I.U.Psy.S.,* Volume 3 (pp. 199–228). Amsterdam: North-Holland.

Geissler, H.-G. (1987). The temporal architecture of central information processing: Evidence for a tentative time-quantum model. *Psychological Research, 49*, 99–106.

Geissler, H.-G. (1990). Foundations of quantized processing. In H.-G. Geissler (Ed.), *Psychophysical explorations of mental structures* (pp. 193–210). Göttingen, Germany: Hogrefe & Huber Publishers.

Geissler, H.-G. (1991). Zeitcodekonstanten—ein Bindeglied zwischen Psychologie und Physiologie bei der Erforschung kognitiver Prozesse? Hypothesen und Überlegungen zu Quantenstrukturen in der Alpha-Aktivität. *Zeitschrift für Psychologie, 199*, 121–143.

Geissler, H.-G. (1992). New magic numbers in mental activity: On a taxonomic system for critical time periods. In H.-G. Geissler, S. W. Link, & J. T. Townsend (Eds.), *Cognition, information processing and psychophysics* (pp. 293–321). Hillsdale, NJ: Erlbaum.

Geissler, H.-G. (1994). Über Möglichkeiten diskret-ganzzahliger Strukturierung in Wahrnehmung und Gedächtnis: Betrachtungen in Sachen Zeitquanten. In D. Dörner & E. van der Meer (Eds.), *Das Gedächtnis* (pp. 19–52). Göttingen, Germany: Hogrefe.

Geissler, H.-G. (1997). Is there a way from behavior to nonlinear brain dynamics? On quantal periods in cognition and the place of alpha in brain resonances. *International Journal of Psychophysiology*, Special Issue, *26*, 381–393.

Geissler, H.-G. (in press). Functional architectures in structural recognition and the issue of seeming redundancy. In C. Kaernbach, E. Schröger, & H. Müller (Eds.), *Psychophysics beyond sensation: Laws and invariants of human cognition*. Hillsdale, NJ: Erlbaum.

Geissler, H.-G., Klix, F., & Scheidereiter, U. (1978). Visual recognition of serial structure: Evidence of a two-stage scanning model. In E. L. J. Leeuwenberg & H. F. J. M. Buffart (Eds.), *Formal theories of perception* (pp. 299–314). Chichester: John Wiley.

Geissler, H.-G., & Puffe, M. (1982). Item recognition and no end: Representation format and processing strategies. In H.-G. Geissler & Petzold (Eds.), *Psychophysical judgment and the process of perception* (pp. 270–281). Amsterdam: North-Holland.

Geissler, H.-G., & Kompass, R. (1999). Psychophysical time units and the band structure of brain oscillations. *15th Annual Meeting of the International Society for Psychophysics*, 7–12.

Geissler, H.-G., Schebera, F.-U., & Kompass, R. (1999). Ultra-precise quantal timing: Evidence from simultaneity thresholds in long-range apparent movement. *Perception and Psychophysics, 6*, 707–726.

Geissler, H.-G., & Kompass, R. (2001). Temporal constraints in binding? Evidence from quantal state transitions in perception. *Visual Cognition, 8*, 679–696.

Goldmeier, E. (1982). *The memory trace*. Hillsdale, NJ: Erlbaum.

Jensen, O. (1998). Oscillatory short-term memory models. Dissertation: Brandeis University.

Jensen, O., Gelfand, J., Kounios, J., & Lisman, J. E. (1998). 10–12 Hz oscillations increase with memory load in a short-term memory task. Internet Note.

Jost, K., Kompass, R., & Geissler, H.-G. (2002). Unpublished experiment.

Klix, F., & van der Meer, E. (1978). Analogical reasoning - an approach to mechanisms underlying human intelligence performances. In F. Klix (Ed.), *Human and artificial Intelligence* (p. 212). Berlin: Deutscher Verlag der Wissenschaften.

Kolers, P. A. (1964). The illusion of movement. *American Sciences, 211*, 98–106.

Kompass, R., & Elliott, M. A. (2001). Modeling as part of perception: A hypothesis on the function of neural oscillations. *Proceedings of the 17th Annual Meeting of the International Society for Psychophysics*, 130–135.

Kompass, R. (2002). Quantal timing: A neural principle and its relation to psychophysics. *Proceedings of the 18th Annual Meeting of the International Society for Psychophysics*, 138–144.

Kompass, R., & Geissler, H.-G. (2001). Quantal timing: An investigation of gamma apparent motion. *Proceedings of the 17th Annual Meeting of the International Society for Psychophysics*, 462–467.

Kristofferson, A. B. (1980). A quantal step function in duration discrimination. *Perception and Psychophysics, 27*, 300–306.

Kristofferson, A. B. (1990). Timing mechanisms and the threshold for duration. In Geissler, H.-G. (Ed., in collaboration with M. H. Müller & W. Prinz), *Psychophysical explorations of mental structures* (pp. 269–277). Toronto: Hogrefe & Huber Publishers.

Kristofferson, M. W. (1972). Effects of practice on character-classification performance. *Canadian Journal of Psychology, 26*, 540–560.

Lisman, J. E., & Idiart, M. A. P. (1995). Storage of 7 ± 2 short-term memories in oscillatory subcycles. *Science, 267,* 1512–1515.

MacKay, D. M. (1963). Psychophysics of perceived intensity: A theoretical basis for Fechner's and Stevens' law. *Science, 139*, 1213–1216.

Maltseva, I., Geissler, H.-G., & Basar, E. (2000). Alpha oscillations as an indicator of dynamic memory operations: Anticipation of omitted stimuli. *International Journal of Psychophysiology, 36*, 185–197.

Petzold, P., & Edeler (1995). Organization of person memory and retrieval processes in recognition. *European Journal of Social Psychology, 25*, 249–267.

Petzold, P., Geissler, H.-G., & Edeler, B. (1999). Discrete clusters of processing time in a verbal item-recognition task. *15th Annual Meeting of the International Society for Psychophysics*, 25–30.

Pöppel, E. (1970). Excitability cycles in central intermittency. *Psychologische Forschung, 34*, 1–9.

Puffe, M. (1990). Quantized speed-capacity relations in short-term memory. In H.-G. Geissler (Ed., in collaboration with M. H. Müller & W. Prinz), *Psychophysical exploration of mental structures* (pp. 290–302). Toronto: Hogrefe & Huber Publishers.

Stroud, J. M. (1955). The fine structure of psychological time. In H. Quastler (Ed.), *Information theory in psychology* (pp. 140–207). Glencoe, IL: Free Press.

Teghtsoonian, R. (1971): On the exponents in Stevens' law and on the constant in Ekman's law. *Psychological Review, 78*, 71–80.

Treisman, M., Cook, N., Naish, P. L. N., & MacCrone, J. K. (1994). The internal clock: Electroencephalographic evidence for oscillatory processes underlying time perception. *The Quarterly Journal of Experimental Psychology, 47A*, 241–289.

Treisman, M., Faulkner, A., Naish, P. L. N., & Brogan, D. (1990). The internal clock: Evidence for a temporal oscillator underlying time perception with some estimates of its characteristic frequency. *Perception, 19*, 705–743.

Vanagas, V. (2001). Quantal timing as a consequence of the anticipatory activity of the nervous system. In E. Sommerfeld, R. Kompass, & T. Lachmann (Eds.), *Fechner Day 2001: Proceedings of the 17th Annual ISP Meeting* (pp. 647–652). Lengerich, Germany: Pabst Science Publishers.

Vanagas, V., Balkelytė, O., Bartusyavitchus, E., & Kirvelis, D. (1976). The quantum character of recognition process in human visual system (in Russian). In V. D. Glezer (Ed.), *Information processing in the visual system* (pp. 26–30). Leningrad: Academy of Science.

von Békésy, G. (1936). Über die Hörschwelle und Fühlgrenze langsamer sinusförmiger Luftdruckschwankungen. *Annalen der Physik, 26*, 554–556.

Chapter 12: Behavioral and electrophysiological oscillations in information processing: A tentative synthesis*

BORÍS BURLE, FRANÇOISE MACAR, and MICHEL BONNET

Abstract

The fine temporal structure of information processing has been an important research topic in both experimental psychology and neuroscience since the 50s. At this time, the α rhythm was thought to act as a temporal device (i.e., a timer) in information processing. Although this conception was subsequently contested, the empirical data obtained at this time and the theoretical conception derived from them may be re-evaluated in light of recent findings. Indeed, over the last 10 years electrophysiological and behavioral researches have suggested the involvement of faster rhythms in information processing. Behavioral and neurophysiological evidence for oscillations in information processing, both sensory and motor, will be reviewed. We will try to synthesize the data based on behavioral techniques, global electrophysiology (electroencephalography, local field potential, multi-unit recording), and analysis of unit activity (membrane potentials oscillations, synaptic determinants of oscillations). A tentative architecture model will be proposed. The main idea is that an internal oscillator, the role of which would be to chunk information flow in packets, paces the information processing system. This idea led us to propose the dual pacing hypothesis, according to which both the information flow and the receptivity state of neurons receiving this information are paced by the oscillator. According to this conception, oscillations would increase signal/noise ratio in processing.

Introduction

For the last 10 years, interest in the fine temporal resolution in information processing have been largely renewed following the work of Gray and Singer (1989) on synchronous oscillations in the cat visual cortex. They reported that when the receptive fields of the recorded neurons were crossed by coherent stimulation, mimicking a single moving object, the two recorded neurons presented a synchronous oscillation. That is, their spikes tended to be precisely synchronized (0 phase shift) and the

* The writing of this chapter was partly made possible thanks to a grant from the "Fondation Fyssen" to the first author. The authors wish to thank Thierry Hasbroucq, Franck Vidal, Andras Semjen and Alexa Riehle for numerous usefull discussions on this topic, and Martin Elton for his comments.

cross-correlogram between the two neurons firing rates presented an oscillation at a frequency around 40 Hz. This result led the authors to propose that synchronous oscillations could be the neurophysiological solution of the so-called "binding problem" in the brain, due to a temporal coding of information (Singer, 1993; see also Engel, this volume, chapter 9). The main idea underlying this proposition is that information is not only coded by the discharge frequency of neurons, but the precise timing of action potentials may also play an important role in information processing. Such a view of brain functioning has often been referred to as the "temporal coding hypothesis". Despite the obvious importance of Gray and Singer (1989) results, the "temporal coding" hypothesis is not new, and such a notion was already present as early as in the 50s. In this chapter, we will first present the arguments that led several authors to propose that a slower rhythm, the so-called α rhythm, may act as a timer in the information processing system (see, e.g., Wiener, 1958). We will then review behavioral data, some old and some more recent, suggesting that information processing is not a continuous process, but is instead periodic. It will then be argued that, when working under attentional constraints, the information processing system is paced, but not driven, by a "central oscillator". This proposition will be related to electrophysiological data on fast oscillations, and a tentative synthesis, the dual pacing hypothesis, will be proposed.

Synchronization versus oscillation

Before going further, it seems important to clarify the goal of this chapter. Gray and Singer (1998) reported synchronous oscillations in the cat visual cortex which they view as reflecting the formation of cell assemblies, in line with Hebb's early proposals (Hebb, 1949) and with Abeles' view (Abeles, 1982) of the cortical neurons as coincidence detectors. However, in both of these conceptions, the important aspect is the synchronization of neural activity: oscillation does not appear as critical. Theoretically, spike synchronization can occur without any need for oscillation, and several authors have argued that neural synchronization, on one hand, and cortical oscillation, on the other hand, should be considered separately, and that they may well reflect different functional mechanisms (see, e.g., Frégnac, Bringuier, & Baranyi, 1994). From a physiological point of view, it is worth noting that synchronization takes place at the level of spikes, whereas oscillations, especially those recorded with "macroscopic" techniques like local field potential or electroencephalography (EEG), are mostly due to synaptic activity (Bringuier, Frégnac, Barnyi, Debanne, & Schultz, 1997). Therefore, if we accept that synchronization is a good candidate for solving the "binding problem", thanks to cell assemblies formation, and if oscillation and synchronization have to be distinguished, the precise role of oscillation, if any, still needs to be established. To disentangle these two phenomena at an empirical level is, however, not straightforward. Indeed, if we can easily conceive neural synchronization without any oscillation (see, e.g., Riehle, Grün, Diesmann, & Aertsen, 1997), the reverse is much more difficult, as oscillatory synaptic afference will tend to synchronize the spiking of neurons by biasing their discharge probability (Murthy & Fetz,

1996). Despite this difficulty, such a distinction remains important, and the remainder of the chapter will focus mostly on possible roles of fast oscillation in information processing.

Temporal coding and α rhythm

When Caton (1875) first introduced an electrode in a monkey brain, recording then probably the first EEG, the most obvious thing he saw was electrical activity oscillation around 10 Hz. Such oscillation, called *α* rhythm, has of course interested psychologists, and a lot of work performed on temporal coding were, more or less explicitly, related to the α rhythm, despite the fact that some behavioral data did not fit with α period (see next section).

The idea that an oscillation around 8–12 Hz was shaping performance had already been proposed by Travis (1929). He was not directly interested in *α* rhythm, but in the physiological tremor, a spontaneous oscillation of the fingers around 10 Hz. At this time, it was thought that the physiological tremor and the *α* rhythm were related.[23] Travis observed that voluntary movement of the fingers, either upward or downward, were more likely to occur during particular phases of the tremor so that the movement will be a continuation of the tremor oscillation. Such a relationship between movement onset and tremor phase was also observed in a reaction time (RT) task. These observations were later replicated by Goodman and Kelso (1983) but were more recently criticized (Lakie & Combes, 2000).

During the 60s, several studies established a relationship between *α* rhythm and simple RT. Among them, the most important were probably Surwillo's publications (Surwillo, 1961, 1963, 1964). In his various studies, he observed positive correlations, often rather strong, between the *α* period and the RT duration. These correlations were obtained either on different populations of subjects (comparison between the mean RT and the mean *α* period) or within subject (Surwillo, 1963). He further extended these data to a choice "go-nogo" task. He observed that the slope of the relationship between RT and α period was steeper in choice than in simple RT, suggesting that this relationship was indeed functional. Woodruff (1975) studied this relationship in a "biofeedback" study. Subjects received feedback on the frequency of their *α* and were asked either to maintain the frequency between predefined values close to the mode, or to speed-up or to slow-down the *α* frequency. For these different *α* speeds, the RT was measured. Shorter RT were obtained when the *α* was speeded-up, whereas subjects were slower when the *α* was slow, compared to the situation in which the *α* remained around its mode.

In the perceptual domain, some data also suggested an involvement of α in information processing. Lesèvre and Rémond (1967) have shown that the amplitude of the visual evoked potential varies as a function of the phase of the α in which the stimulus is presented. These data were in agreement with the idea that the α reflects an oscillation of the cortical excitability. More direct arguments for a role of α rhythm in

---

[23] Travis himself will later invalidate this proposition (Travis & Cofer, 1937)

perceptual processes come from a study by Varela, Toro, John, and Schwartz (1981). They were interested in the perception of simultaneity. When one presents two visual stimulations spatially separated by only a few degrees, if the temporal interval between the two stimulations is small, the two stimulations are perceived as simultaneous whereas if the temporal separation is long enough, the subjects perceive the two stimulations as sequential. Varela and colleagues presented the stimuli at different phases of the α (positive versus negative part) and observed that the probability to perceive the two stimuli as simultaneous versus sequential systematically varied as a function of the phase of the α in which the stimulation was presented. This provides strong arguments for the existence of perceptual episodes, inside which it is not possible to segregate the different pieces of information.

Notwithstanding the results presented above, the idea that the α is involved in information processing has been criticized. For example, Boddy (1971) did not observe the correlation between α period and RT described by Surwillo. He therefore rejected the idea that α could affect information processing, and proposed that the correlations obtained by Surwillo can be explained in terms of the arousal level of the subjects, affecting simultaneously the α period and the RT, without implying that these two variables are directly correlated.[24] Treisman (1984) addressed even more directly the question as to whether the α rhythm may act as a timer for the information processing system. He compared, on a subject by subject basis, the performance in time estimation and the α period. Whatever the comparison performed (mean frequency, variability etc.), no clear relationship could be found. Therefore, Treisman (1984) concluded against the α rhythm acting as a timer.

Apart from the studies on the α rhythm, several works, coming mostly from experimental psychology and/or psychophysics, suggested, already in the 60s, that information processing is not continuous, but is instead periodic (see also Geissler & Kompass, this volume, chapter 11). Interestingly, however, as we will see, several behavioral results already suggested much faster rhythms as candidates to be involved in information processing.

Periodicity in human information processing

Integration windows in sensory processing

Psychologists interested in simultaneity perception evidenced that the environmental information is perceived through temporal windows (Schmidt & Kristofferson, 1963; Hirsh & Fraisse, 1964), also sometimes called "perceptual moments" (Stroud, 1955). This general idea has been delineated under different variants (see Bertelson, 1966; Harter, 1967, for reviews), with some discussions on whether such moments are associated with moving windows or discrete ones (Allport, 1968). A

[24] This addresses the problem of how to interpret a correlation. If A is correlated with B, there are (at least) three possible cases: (1) A is the cause of B, (2) B is the cause of A, or (3) a third factor, C, correlates with A and B. In the latter case, a residual correlation between A and B can be erroneously interpreted.

formal analysis of the perceptual windows hypothesis allowed Shallice (1964) to show that such a discontinuity in the perception is the most economic mechanism to detect changes in the environment. This analysis was in line with Wiener's very early proposals that some "internal clocks," acting as temporal referent in the central nervous system, were necessary (Wiener, 1958). For Wiener, the role of such internal clocks could be to chunk the information flow in discrete events. A strong argument for a non continuous perception was provided by Latour (1967). He presented to his subjects two light flashes in succession. The delay t between the two flashes was varied from 175 to 225 ms. He observed that the probability to detect the two flashes was a periodic function of t, that is, the probability of perceiving the two flashes was oscillating. Such an oscillation was still present up to 225 ms between the two lights, excluding a possible retinal persistence effect. The observed periodicity had a 27 ms period, that is in the γ range (see later).

More recently, Andrews, White, Binder, and Purves (1996) introduced new arguments in favor of a discontinuity in the visual perception, by replicating and extending results obtained by Sherrington (1906) about one century ago (cf. Sherrington, 1947). Sherrington was interested in flicker fusion frequency, that is the frequency above which a series of flashes is perceived as a continuous stimulation. Two experimental conditions have been used: the stimulation of the two eyes was either synchronous or asynchronous (in anti phase). A 20 Hz stimulation, synchronous for both eyes corresponds to a stimulation of the subject every 50 ms, whereas an asynchronous stimulation at 20 Hz on both eyes corresponds to stimulation of the subject every 25 ms, thus equivalent to the double frequency (40 Hz). Sherrington's predictions were that the asynchronous binocular fusion frequency should be half the synchronous fusion frequency. Indeed, if one assumes that the fusion frequency is, let's say, 50 Hz, thus a 25 Hz asynchronous stimulation should be equivalent to a 50 Hz synchronous stimulation. However, this is not the case: The fusion frequency is the same for the synchronous and asynchronous stimulation. These data suggest that fusion is not (only) of peripheral (retinal) origin, but instead that it occurs at a level in which the binocular integration is already done, that is at least at the V1 level.[25] In order to go further, Andrews et al. (1996), in a second phase, presented on each trial a sequence of synchronous oscillation followed by a asynchronous one. The stimulation frequencies were varied from trial to trial. The subjects had to decide whether the first stimulation (synchronous one) was identical to the second one (asynchronous). At relatively slow frequencies, the equality was perceived when the asynchronous stimulation (e.g., 5 Hz) was half the synchronous one. But this matching to the double frequency disappeared for higher frequencies: A 25 Hz synchronous stimulation was perceived as equal to a 25 Hz asynchronous stimulation. The authors concluded that: "[...] we ordinarily sparse visual input into temporal episodes. If sequential flashes to alternate eyes fall into successive episodes, then our perception is of a union of the dichotic input, as indeed occur at very low frequencies of presentation. If however, sequential stimuli fall within one episode, the information is conflated [...]." Purves, Paydarfar, and Andrews (1996) proposed that such a "sampling" of the

[25] We are grateful to Eric Castet for discussions we had on this question.

visual information could also account for the “wagon wheel illusion,” that is, for some speed, our perception of wheel rotating in the sense opposite to the motion. This illusion, very clear in movies, where images are sampled, is also present in continuous natural lighting conditions, which suggests that, at least in this case, it is indeed the visual system that samples and splits up the information flow (Purves et al., 1996).

Oscillations in motor behavior

Although not so important, or perhaps not so well documented, periodic phenomena are not restricted to sensory processing, but also appear on motor performances. Such a periodicity has first been studied on tasks where a precise control of timing was required, like tapping at a stable rate (Wing & Kristofferson, 1973a, b; Vorberg & Hambuch, 1984; Collyer, Broadbent, & Church, 1992, 1994). However, periodicity in motor actions is not only observed in rhythmic behavior, but also appears in more continuous movements. For example, Vallbo and Wessberg (1993) have studied fingers displacements in slow visual target tracking. They observed that displacements were not smooth, but instead discontinuous and periodic with a frequency around 8 to 10 Hz. This periodicity, not always visible on raw displacement traces, clearly appears on the various derivatives of the position, that is speed, acceleration and jerk. McAuley, Farmer, Rothwell, and Marsden (1999) have extended these data to ocular movements. They observed that ocular movements were also periodic in a tracking task, with the same frequency. Furthermore, when subjects had to simultaneously track a moving target with both eyes and a finger, there was a high coherency between the two signals, suggesting a stable phase relationship between eyes and finger movements. These data were interpreted as reflecting the involvement of a central temporal system, common to the various motor systems, which organizes the timing of movements.

Periodicity in reaction time

A large part of studies on mental activities relies on a chronometrical investigation, initiated in the 19th century by Donders (1868; see, e.g., Luce, 1986; Posner, 1978; Sanders, 1998, for reviews). In the 60’s, several authors, using RT as a measure of performance, analyzed precisely the distribution shape. The first important characteristic, not really relevant here, is that the distribution is not Gaussian, but instead asymmetric, with a right tail longer than the left one (see, e.g., Hohle, 1965). The second point is more relevant: the distribution seemed not to be unimodal, but instead multimodal, with a constant period between the peaks. In other words, the distributions were periodic, with a period of 25 ms (i.e., 40 Hz) between peaks (Latour, 1967; Harter & White, 1968). Harter and White (1968) also reported a similar periodicity in the EMG activity of the muscles involved in the response. These studies, and some other similar ones (Pöppel, 1996) have been severely criticized because of inadequate signal processing methods (Allan, 1995; Vorberg & Schwartz, 1987; Vroon, 1974). The criticisms were mostly based on the fact that the distributions were split and grouped in several classes of equal length, and in case of few data, this can lead to artificial variation in the number of trial per classes. Nevertheless, more recently, with more adequate methods and more rigorous statistical analysis, Dehaene (1993) has also found a clear periodicity in RT distributions. In Dehaene's (1993) study, four tasks were used: a visual feature task, a visual conjunction task, an auditory feature task and an auditory conjunction task. Two results are worth noticing: first, as presented above, Dehaene confirmed the existence of a periodicity in RT distribution, and, second, the period of the oscillation was longer for the tasks leading to longer mean RT. Dehaene therefore argued that responses are emitted after a relatively constant number of oscillation cycles, and hence that the differences in RT obtained between the four tasks are mostly due to variations in the oscillation period rather than to an increase in the number of cycles.

The functional interpretation of these oscillations necessitates some explanation. Such a periodicity means that the probability for a response to be given is not regular in time, but that there exists some particular moments, after the stimulus presentation, where the response to this stimulation is more likely to occur, and hence that at other moments, the response is less likely to occur. In other words, there exists some preferred interval (separated by about 25–30 ms) after the stimulus presentation where the responses are grouped. Because of the oscillations in EMG activities, Harter and White (1968) hypothesized a motor gate, opening periodically, which allows the response to be emitted at determined moments only. Note that in this case, the processing of information is not necessarily periodic, only the motor output is. On the opposite, one can however suppose, in line with Harter (1967) and Latour (1967), that although the information flow received from the environment is continuous, in order to avoid a saturation, the perceptual system samples the information and chunks it in "information packets." This idea is the core of the model developed by Dehaene (1993) to account for periodicity: He proposed that perceptual information is transmitted by packets at regular moments in time. For each packet a decision is made, either to respond immediately, or to wait for the next packet in order to transmit more complete information to the response system (see Figure 1). We shall return on that point later.

Periodicity has also been observed in saccadic RT. Indeed, if one introduces a gap between the disappearance of the fixation point and the response signal onset, saccade latency distributions become bimodal with the two modes being locate around 100 and 150 ms, respectively. Saccades centered around 100 ms are called express saccades. Two interpretations have been proposed to explain this bimodality: either the participation of different anatomical loops, one for each type of saccades (express or normal, see Fisher & Weber, 1993, for a review), or the existence of some oscillations in saccade triggering thresholds (Kirschfeld, Feiler, & Wolf-Oberhollenzer, 1996).

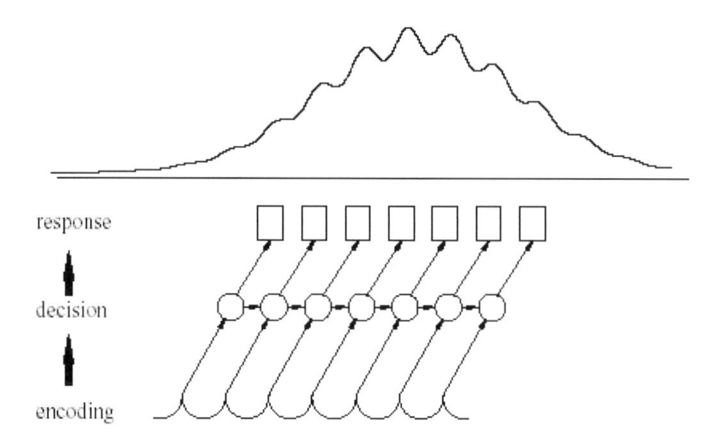

Figure 1: Schematic representation of the information processing system architecture proposed by Dehaene (1993). Adapted from Dehaene (1993).

"Quantal" models of information processing periodicity

In order to interpret the results on simultaneity perception, and on time perception, Kristofferson (1967, 1976, 1980, 1984) proposed a quantal conception of information processing. To study the discrimination of durations, he developed a model in which a race occurs between two responses (shorter versus longer). From this model and from the probability density function of the responses, he derived the value of the time quantum in duration estimation (see Kristofferson, 1984, for more detailed explanations). He observed that the duration of the quantum depends on the duration to be estimated. More importantly, the different values of the quantum obtained were 12; 24; 48 and 96 ms. As we can see, the various values of the quantum follow a doubling rule, and it is worth noting that they correspond to about 80; 40; 20 and 10 Hz frequencies.

Another influential quantal model has been proposed by Geissler (1987), based on a review of the literature (see also Geissler & Kompass, 2001). This conception is presented in the chapter by Geissler and Kompass in this volume, and interested readers are referred to this chapter (chapter 11).

A temporal oscillator in information processing?

Periodicity is often, and naturally, conceived as the result of an oscillator (or pacemaker) activity. For example, in his temporal information processing model, Treisman (1963) proposed that a temporal oscillator sending pulses at regular intervals serves as a time base. The idea that an oscillator is involved in time estimation is present in a large number of temporal processing models (Gibbon, Church, & Meck, 1984) and of rhythmic motor production models (Cattaert, Semjen, & Summers,

1999; Vorberg & Hambuch, 1984; Wing & Kristofferson, 1973b, 1973a). All these models dissociate two levels, one concerning a central oscillator, and another one concerning motor execution, even if they sometimes diverge on the precise role of the oscillator (Billon, Semjen, Cole, & Gauthier, 1996). They are clearly different from the so-called "dynamic models" which do not dissociate these two levels (see, e.g., Kelso, Holt, Rubin, & Kugler, 1981).

In the 90's, Treisman and collaborators (Treisman, Faulkner, Naish, & Brogan, 1990; Treisman, Faulkner, & Naish, 1992; Treisman, Faulkner, Naish, & MacCrone, 1994) proposed a more precise model for the oscillator, and more importantly, a protocol to study its existence and its frequency. The model consists of two parts: a temporal oscillator (TO) and a calibration unit (CU). The TO is modeled by Treisman et al. as a set of three neurons, each neuron sending activation or inhibition to the other ones. The TO pulses at a stable frequency *Fo*. The produced pulses are the inputs of the calibration unit, which multiplies the input frequency it receives by a calibration factor *Cf* to produce its output frequency *Fp*, with $Fp = Cf \times Fo$. This mechanism permits a stable frequency, which can be easily adjusted if necessary: suppose that, after having programmed a movement, something happens in the environment which necessitates to speed-up the execution, an increase in *Cf* will allow a faster execution, without any lost of coordination. Treisman and colleagues further assumed that, although the TO must be protected from outside stimulation in order to keep a stable frequency, a strong enough periodic sensorial input may drive it. If the external and internal frequencies are close enough, a phase locking will occur, which will slow down the TO if the external stimulation is slower than *Fo*, and speed it up in the opposite case. We thus obtain a characteristic pattern on the oscillator frequency as a function of the external stimulation (Figure 2).

From this interference pattern, Treisman et al. concluded that it is possible to estimate the value of *Fo*. This model has been tested in duration estimation and production tasks (Treisman et al., 1990). The obtained estimations of the oscillator frequency were 24.5, 37.3 and 49.5 Hz, which correspond approximately to 2, 3 and 4 time 12.4 Hz. The next step was to evaluate whether time estimation and timing of movement rely on the same mechanism. Treisman et al. (1992) extended these results to sensorimotor activities like reaction time and typing. In these experiments, the estimated value was 49.8 Hz. Even if one can question the validity of such precise estimation, Treisman's method remains useful to study the presence and the role of such an oscillator in information processing.

Following Treisman and collaborators' data, Burle and Bonnet (1997) addressed the question as to whether such an oscillator affects equally all the processing stages, or only some specific processes. Indeed, in a RT situation, in order to elicit the appropriate response, the information conveyed by the stimulus has to go through various processes. A minimal decomposition relies on three stages of information processing: stimulus perception, stimulus-response association and response elaboration (Proctor, Reeve, & Weeks, 1990). To address which processes are paced by the pacemaker, Burle and Bonnet (1997) combined Treisman's protocol with the additive factor method (AFM, Sternberg, 1969a, 1998). The logic of the AFM can be summarized as follows (see Sanders, 1998, for a recent review and a more complete description): If two factors affect different processing stages, their joint effects on the RT is equal to the sum of their individual effects; in other words, the two factors have additive effects on the performance. By contrast, if two factors affect at least one stage in common, they should interact overadditively, that is, their joint effect should be higher than the sum of their individual effects. By manipulating factors known to affect the sensory, decisional and motor stages on one hand, and the oscillator frequency, thanks to Treisman's protocol on the other hand, it was possible to check for interaction between the oscillator speed and the manipulated factors. First, the study confirmed the existence of nonlinearity in the RT as a function of external stimulation frequency, as predicted by Treisman's model. The interference between the external frequency and the internal oscillator was located around 20 Hz, suggesting that the TO frequency has an harmonic at this value. Second, the amplitude of the interference was higher for long RTs than for short ones, confirming that the oscillator paces some processes within the RT. Nevertheless, because of the small size of expected effects, the data on which information processing stages are affected were inconclusive.

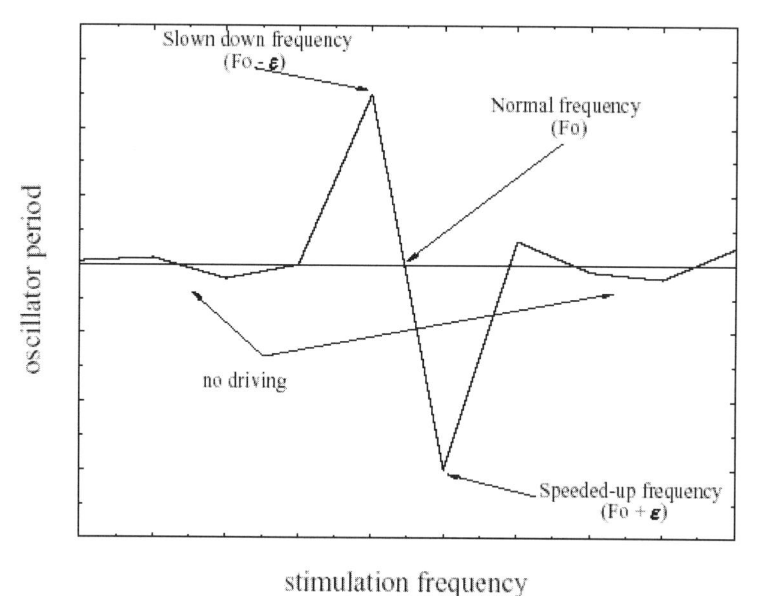

Figure 2: Representation of a driving effect on an hypothetical oscillator as a function of the external stimulation frequency (abscissa). The internal oscillator, which pulses at a given frequency *Fo*, will be driven by the periodical stimulation provided that the two frequencies are close. If the stimulation frequency is lower than the internal one, the effect is a slowing down of the internal rhythm, and hence an increase in the processing time. A speeding up will occur if the stimulation frequency is just above *Fo*. Adapted after Treisman et al. (1990).

In a next step, Burle and Bonnet (1999) evaluated the possibility that such an oscillator could pace the transmission of information between information processing stages. They first replicated their previous results, and hence those of Treisman and collaborators, as they obtained the interference for the same values as before. Their results also suggested that the oscillator is involved in the transmission of information between processing stages.

In the two above reported experiments, Burle and Bonnet (1997, 1999) used slightly different values for the external stimulation frequencies. This allows a test of the robustness of the results. Figure 3 presents the combined data obtained from the two experiments. As we can see, they combine very nicely, strengthening the reliability of the "interference pattern."

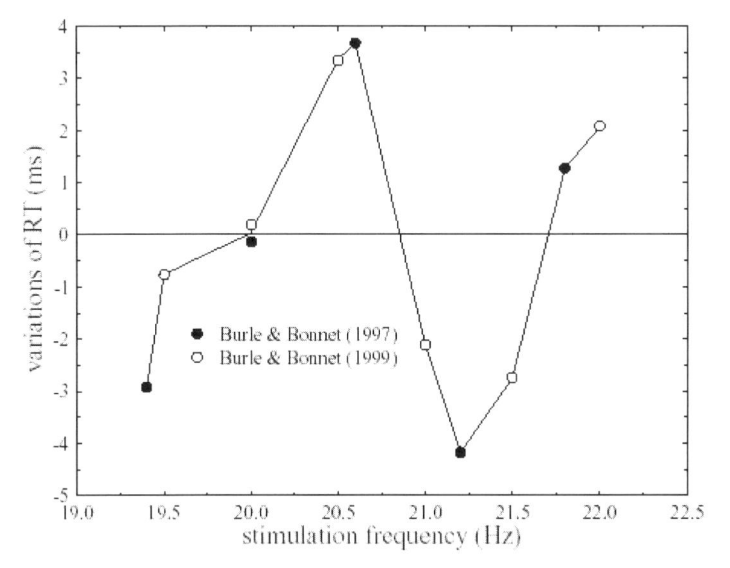

Figure 3: Comparison of the data obtained in two different experiments, where the stimulation frequencies were slightly different. The two data sets combine very nicely, and follow the pattern predicted by Treisman et al.'s model.

Treisman's protocol also allowed Burle and Bonnet (2000) to test a neurophysiological model of short-term memory. Lisman and collaborators (Lisman & Idiart, 1995; Jensen & Lisman, 1998) have proposed that a neural network using θ (5–12 Hz) and γ (20–60 Hz) oscillations can maintain about seven items in memory. The basic idea is that each item is stored in one γ cycle, and that the response can be emitted only at the end of the θ cycle. This model nicely accounts for two major findings in short-term memory, namely the memory span (Miller, 1956) and the high speed scanning (Sternberg, 1966, 1969b). In his famous experiments, Sternberg (1966) presented a list of items (for example, digits) to be memorized by the subjects. The number of items in the list was varied. He then presented one digit, and the subject had to decide as fast as possible whether the presented digit belonged to the memorized list or not. Sternberg (1966) observed that the RT increased linearly with the length of the list, and that the slope of the relationship was the same when the digit belonged or not to the list. Based on these results, Sternberg argued that memory search is serial and exhaustive. In Jensen and Lisman's model (Jensen & Lisman, 1998), the slope of this relationship is determined by the γ oscillation period. Therefore, if Treisman's protocol does entrain the oscillation period, then we may not only expect an effect of stimulation frequency on the mean RT (as obtained in the previous experiments), but also on the value of the slope relating RT to the number of items in the list. This is what Burle and Bonnet (2000) obtained, providing arguments both for Jensen and Lisman's model and for the involvement of an oscillator in information processing.

In conclusion, the involvement of an oscillator in the information processing system has been widely documented in the behavioral literature. Previous attempts to relate those results to electrical cortical oscillations failed, mostly because such oscillations in behavior were thought to be related to the α rhythm (see above). However, such a relationship can now be re-evaluated in light of recent data on faster rhythms.

Fast electrical oscillation in human information processing

It is not the goal of this section to review the results obtained on γ oscillation, and interested readers are referred to excellent reviews on this topic (Singer, 1993). However, most of these reviews were mostly concerned with perceptual processes (see MacKay, 1997, for an exception). We will first present early observations on the γ oscillation, along with their functional interpretations. Thereafter, the role of oscillation on motor preparation will be briefly reviewed, and finally the relation between γ oscillation and attentional processes will be outlined.

First observations

Independent of the criticisms on the α rhythm (see above), one may also wonder why researches were so focused on this rhythm, as several results provided arguments for the involvement of faster rhythms. For example, in the very same paper where he reported oscillations in perception and in RT distribution (see above), Latour (1967) reported that "[...] while the subject is performing his task, his α activity is suppressed and a rapid rhythm occurs [...]." This fast rhythm is reproduced in the Figure 2 of Latour (1967), and if one computes the period of this rhythm, one obtains about 25 ms, that is 40 Hz. However, very little, to say the least, is said about the possible role of this rhythm in information processing. In the 70s, Sheer (1970, 1976) also reported γ oscillations during information processing. These were, however, not interpreted as reflecting the formation of object representations (Bertrand & Tallon-Baudry, 2000), but instead as reflecting a global state of the system, that the authors called "focussed arousal." In line with this view, Bouyer, Montaron, and Rougeul (1981) reported that when a cat was attentively observing a mouse, γ oscillations were clearly visible in the corticogram of the cat, and these oscillations disappeared when the mouse was hidden. This observation led the authors to interpret such fast oscillations as reflecting a "focused attention" state.

Oscillation and motor preparation

Cortical oscillations related to motor aspects are not as well documented as the perceptual ones, despite the fact that the first observations of periodicity in behavior were motoric (Travis, 1929; Travis & Cofer, 1937). However, some of these data shed important lights on the role that oscillation may have on information processing.

During the preparatory period of a reaction time task, Roelfsema, Engel, König, and Singer (1997) observed synchronous oscillations through several brain areas of cats: After the onset of a stimulus on a screen indicating the beginning of a trial, the cats had to respond to a change in orientation of the stimulus which could be of 90 or 180°. During the preparatory period, between the trial onset and the orientation change, the occipital, parietal and frontal areas presented an oscillation of their local field potential around 20 Hz. These oscillations were synchronous, that is, there was no phase delay. This synchrony disappeared after the orientation change, that is during the execution phase of the RT. The authors interpreted this synchrony as an experimental evidence for the visuo-motor integration based on oscillations.

However, the association between preparatory activity and information integration is not self evident. Indeed, two types of preparations are usually dissociated (Requin, Brener, & Ring, 1991): event preparation and temporal preparation. In case of event preparation, the subject prepares himself specifically to one or several parameters of the incoming stimulus-response association, for example the reception of a red (versus green) signal, or to the emission of a right (versus left) response. In this case, one usually assumes that if the subject knows in advance some of the parameters of the forthcoming movement, some programming of these parameters can be performed during the preparatory period (Rosembaum, 1980; Bonnet, Stelmach, & Requin, 1982; Vidal, Bonnet, & Macar, 1995). By contrast, the temporal preparation is an aspecific process by which the subject tries to synchronize his preparation state with the occurrence of the stimulus onset (Gottsdanker, 1975). In this case, no programming operation can be done during the preparatory period.[26] Thus, even if these two kinds of preparation lead to shorter RT, for event preparation, this shortening is due to the fact that some operations are performed before the appearance of the response signal, whereas in case of temporal preparation, the RT shortening is due to the fact that the operations performed within the RT period are performed faster because the system is in an optimal state to process the incoming information.

In Roelfsema et al.'s study described above, it is not possible to dissociate these two types of preparations. It is therefore very difficult to know whether the oscillations reflect a event or temporal preparation. If the oscillations reflect temporal preparation, they do not express the binding of visual and motor components of the

---

[26] Aspecific (or temporal) and specific (or event) preparation have been proposed to be the motor equivalent of attentional processes in perception, namely sustained and focused attention, respectively (Requin et al., 1991; Coquery, 1994; Brunia, 1999).

task, as proposed by Roelfsema et al., but more likely a non-specific state of the system, allowing an optimal performance of the operations required by the task (Steriade, 1993). In order to decide between these two positions, one may vary the amount of information given to the subject during the foreperiod. This was done by Nashmi, Mendoça, and MacKay (1994) who opposed these two possible roles of oscillations in humans by contrasting simple and choice RT. They analyzed EEG activity over motor and somesthetic areas contralateral to the responding arm. In order to obtain a reliable estimation of the local activity and to separate motor and somesthesic areas, they computed the surface laplacian by mean of the source derivation method (Hjorth, 1975; McKay, 1983). They observed an increased γ activity for the two RT tasks compared to a rest condition. However, no difference showed up between the choice and simple RT, suggesting an involvement of γ activity in temporal (non-specific) preparation rather than event preparation. It is worth noting that during the preparatory period of a RT task the excitability of the cortico-spinal tract decreases (see Bonnet, Requin, & Semjen, 1981, for a review), and that this decrease does not depend on the amount of information given to the subject (Hasbroucq et al., 1999). Such a decrease has been interpreted as reflecting an increased sensitivity of the motoneuronal pool to supraspinal commands (Requin et al., 1991; Schieppati, 1987). In other words, this decrease in excitability may reflect a mechanism that increases the signal/noise ratio in the processing of the motor command. Now, as γ oscillations are mostly due to inhibitory interneurons (see below), the decrease in excitability and the appearance of γ oscillations during the preparatory period may both reflect the very same mechanism whose role would be to increase signal/noise ratio in information processing.

In other conditions, Nashmi et al. (1994) compared several tasks: simple circle drawing, signature drawing ("automatic" tasks), opposed to accurate circle drawing and mental simulation of the latter task ("attentional" tasks). As they observed more γ activities in "attentional" than in "automatic" tasks, they concluded that γ oscillations likely reflect attentional states, rather than the emergence of a motor representation by temporal binding. Several other authors have tried to apply the binding conception at the motor level (Sanes & Donohogue, 1993; Murthy & Fetz, 1996; Aoki, Fetz, Shupe, Lettich, & Ojemann, 1999; see MacKay, 1997, for a review), and all the studies concluded more or less negatively.

γ oscillations and attentional processes

The relationship between γ oscillations and attentional processes has been outlined very early. As previously mentioned, Bouyer, Montaron, and Rougeul (1981) interpreted the high frequency bursts in cat fixating a mouse in terms of attentional modulation. However, following Gray and Singer (1989)'s results, and because of the assimilation between neural synchronization and oscillation, this view became rather marginal, and the notion that γ oscillations could be involved in the formation of representations gained interest. Nevertheless, recently, more and more arguments came to support the notion that γ oscillations are more likely to reflect attentional states (Murthy & Fetz, 1996; MacKay, 1997; Müller & Gruber, 2001; Herrmann & Knight, 2001). For example, the amplitude of the γ response to stimulation is increased when attention is directed towards this stimulation when compared to a condition without attention. Such an increase has been observed in both auditory (Titinen et al., 1993) and visual (Fries, Reynolds, Rorie, & Desimone, 2001) modalities. The involvement of the γ oscillation in attentional processes is also supported by anatomical considerations: The thalamus has often been proposed to be involved in attentional processes (Crick, 1984; LaBerge, 1995; Brunia, 1999). In rats, Pinault and Deschênes (1992) observed reticulo-thalamic cells that "discharged like clocks" (Pinault & Deschênes, 1992, p. 245). These neurons induce inhibitory rhythmic modulation of the thalamo-cortical tract. High frequency rhythmic cells have been observed in several parts of the thalamus, both sensory (Nuñez, Amzica, & Steriade, 1992; Canu, Buser, & Rougeul, 1994; Barth & MacDonald, 1996) and motor (Steriade, Curró-Dossi, Paré, & Oakson, 1991; Timofeev & Steriade, 1997). Therefore, not only does the thalamus innervate all the cortical regions, but furthermore, its various nuclei own high frequency neurons.

The involvement of the thalamus in the genesis and/or the control of fast cortical oscillations is supported by studies coupling recordings at several levels (cortical, thalamic, and even retinal). Steriade and Amzica (1996) recorded oscillatory activity in the parietal cortex (area 5 and 7) simultaneously with the thalamus (centrolateral intralaminar nucleus). They observed synchronous fast (30–40 Hz) rhythms between the two structures. Another line of argument for the involvement of the thalamus, if not in the genesis, at least in the control of fast cortical oscillations comes from stimulations of thalamic nuclei. It has been shown that stimulation of the corresponding thalamus nuclei induces a potentiation of the cortical γ oscillations. This has been shown for somesthesy (Canu et al., 1994), audition (Barth & MacDonald, 1996), and vision (Munk, Roelfsema, König, Engel, & Singer, 1996).

A functional role for the temporal oscillator

The above discussion provides arguments for the involvement of a temporal oscillator in information processing. However, two major questions remain, namely, how does it operate, and what is its role?

Coding information by "packets"

As we have seen above, to account for the periodicity in RT distributions, Dehaene (1993) proposed a temporal organization of information processing based on "packets" of information. According to this view, the environment is perceived periodically (Latour, 1967; Harter, 1967), and after each "packet" a decision is made, either to transmit the results of the processing to the next stage, or to wait and integrate the next packet, in order to give a more complete information to the decision stage. Packets of information are thus transmitted from stages to stages until response execution, explaining why responses are more likely to occur at periodic moments after the stimulus onset. This notion of "packet" of information is also present in neurophysiology. For example, Bringuier et al. (1997) recorded the activity of V1 neurons in cats and kittens. They observed that oscillations are rather seldom in the firing of unitary neurons. However, when the intra-cellular recordings are performed with KCl electrodes, that allow to hyperpolarize the cell, subliminal oscillations of the membrane potential show up. Based on those observations, the authors proposed that the visual information is transmitted by packets, thanks to such oscillations. Furthermore, according to the authors, those oscillations are due to a GABAergic afference on those neurons, that is to say, the membrane oscillations are of synaptic origin. The inhibitory nature of the oscillations is a remarkable characteristic. Indeed, it is established in almost all studies, and is also present in simulation studies (Lytton & Sejnowsky, 1991).

This notion of information packets is also present in Lisman (1997). As synaptic transmission is not reliable, Lisman (1997) proposed that information is not coded by single spikes, but instead by bursts, or train, of spikes, as such trains have been shown to increase synaptic transmission reliability. These bursts of spikes might be functionally equivalent to the packets proposed by Bringuier et al. (1997).

Another argument for a coding of information in packets comes from studies of the cortico-muscular communication. Several studies have evidenced that the cortico-spinal command is not continuous, but instead periodic, that is, it is constituted by brief repetitive impulses between 20 and 50 Hz (Conway et al., 1995).

The dual pacing hypothesis

Coding information in packets means to give a temporal format to the continuous flow of information. Such a format is potentially useable by the information processing system which has to deal with such information. Indeed, it could "recognize" the information (i.e., the relevant messages) in the noise, thanks to such a format. The temporal oscillator could precisely play such a role. Following Jefferys, Traub, and Whittington (1996), we propose that, by inhibiting periodically the pyramidal neurons, the oscillator allows the spiking of neurons only at well defined moments, which would induce a temporal coding in packets (Bringuier et al., 1997). In addition to such a role in sending the information, the oscillator may also be involved in the reception of this information: the periodical inhibition produced by the oscillator could also be sent to the downstream neurons which are the receptors of the packets of information. Such a subliminal inhibition, already described by several authors (Bringuier et al., 1997; Traub, Whittington, Standford, & Jefferys, 1996; Lytton & Sejnowsky, 1991), might induce an oscillation in the membrane potential, and hence periodic changes in neurons excitability. Therefore the downstream neurons may oscillate between receptive and non receptive (inhibited) states. Now, because of the dual pacing, of both the information packets transmission and the excitability states of the downstream neurons, the information from the upstream neurons will always arrive during the receptive states of the downstream ones. This insures reliable transmission of information. By contrast, the cortical noise, that is the irrelevant activities, which is, by definition, randomly distributed is equally likely to occur during the receptive and the non-receptive (inhibited) phases. When occurring in the non-receptive phases, the noise would have almost no effect on the current processing. Such a mechanism could therefore improve the signal/noise ratio in information processing (see Figure 4).

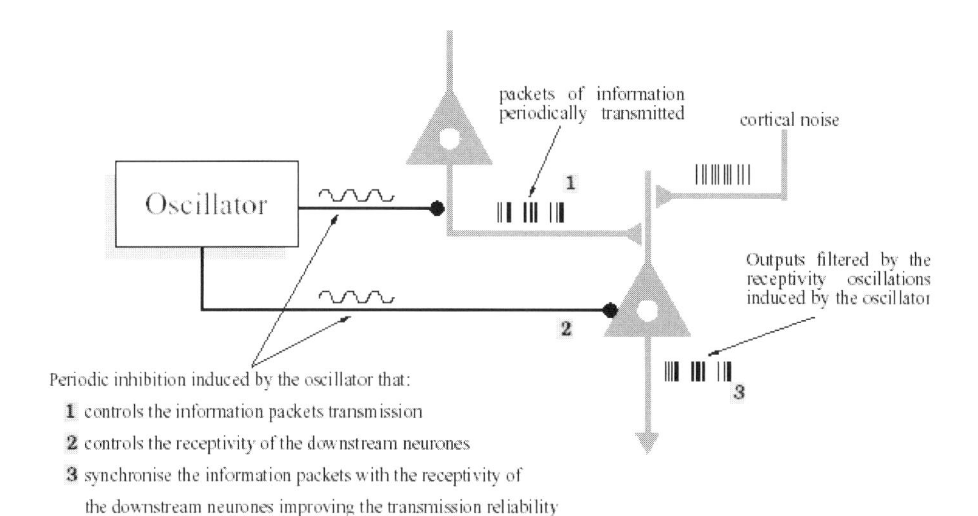

Figure 4: The dual pacing hypothesis: The temporal oscillator simultaneously controls the moment when the upstream neurons can fire, and the receptivity state of the downstream neurons. Such a dual pacing, allowing information transmission in packets, insure that the packets will always arrive during the excitable phases of the downstream neurons. On the contrary, the cortical noise, that is, the irrelevant message for the current processing, will arrive with equal probability during the excitable and inhibited phases. The impact of the noise will be reduced in the latter case, increasing the signal/noise ratio.

Several things are to be noted: First, this conception follows Jefferys et al. (1996) who stated that "[...] gamma oscillation is not proposed to represent information itself, but rather to provide a temporal structure for correlations in the neurons that do encode specific information [...]" (p. 203). Second, there is another way, certainly much more economic, to increase the signal/noise ratio: simply to accumulate more information and to sum information over time. The noise, randomly distributed, will cancel out, whereas the information, more systematic, will consolidate. It is certainly this mode that the central nervous system is using most of the time, at least when not under temporal pressure. However, when such a mode of functioning is no more usable, either because of time pressure, or because the signal/noise ratio is intrinsically bad, the coding of information in packets, and a dual pacing mechanism, might be preferable. We rejoin here the idea that oscillations are related to attention.

As we have seen, the thalamus seems to play an important role, both in the control of cortical oscillations, and in attentional processes. The interconnections between attention, cortical oscillation and the thalamo-cortical pathway are summarized in Figure 5, with a, nonexhaustive, list of references for each connection.

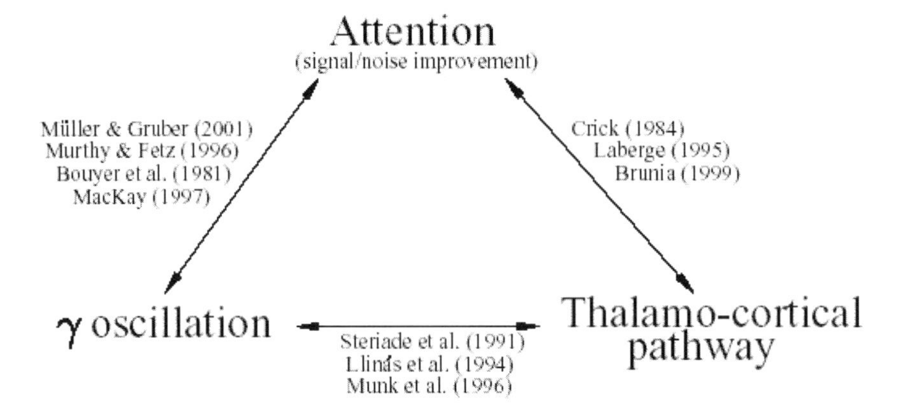

Figure 5: Schematic representation of the relationships between attention (the function), cortical oscillations (the mechanism) and the thalamo-cortical tract (the structure), with the main, nonexhaustive, references on such relationships.

Conclusions

It seems that, when working under attentional constraints, the information processing system is paced by an oscillator. The role of such an oscillator might be twofold: first, to code the information flow in packets, and second, to pace the excitability of the neurons receiving this information. This dual pacing might insure a reliable transmission of information. The oscillator could consist of inhibitory interneurons (Llinás, Grace, & Yarom, 1991), or small networks of interneurons (Jefferys et al., 1996), driven by thalamic afferences (Llinás, Ribary, Joliot, & Wang, 1994). These thalamic afferences might insure a close synchronization of the oscillations over separate cortical areas. Note that, according to the above propositions, oscillation is not necessary to information processing, it is simply useful, especially when working under time pressure, or when the task is intrinsically difficult.

References

Abeles, M. (1982). Role of cortical neuron: Integrator or coincidence detector? *Israelian Journal of Medical Science*, *18*, 83–92.

Allan, L. (1995). Psychological time: Continuous or discrete? In C.-A. Possamaï (Ed.), *Fechner day 95* (pp. 133–138). Cassis, France: International Society for Psychophysics.

Allport, D. (1968). Phenomenal simultaneity and the perceptual moment hypothesis. *British Journal of Psychology*, *59*, 395–406.

Andrews, T., White, L., Binder, D., & Purves, D. (1996). Temporal events in cyclopean vision. *Proceedings of the National Academy of Sciences USA, 93*, 3689–3692.

Aoki, F., Fetz, E., Shupe, L., Lettich, E., & Ojemann, G. (1999). Increased gamma-range activity in human sensorimotor cortex during performance of visuomotor tasks. *Clinical Neurophysiology, 110*, 524–537.

Barth, D., & MacDonald, K. (1996). Thalamic modulation of high-frequency oscillating potentials in auditory cortex. *Nature, 383*, 78–81.

Bertelson, P. (1966). Central intermittency twenty years later. *Quarterly Journal of Experimental Psychology, 18*, 153–162.

Bertrand, O., & Tallon-Baudry, C. (2000). Oscillatory gamma activity in humans: A possible role for object representation. *International Journal of Psychophysiology, 38*, 211–223.

Billon, M., Semjen, A., Cole, J., & Gauthier, G. (1996). The role of sensory information in the production of periodic finger-tapping sequences. *Experimental Brain Research, 110*, 117–130.

Boddy, J. (1971). The relationship of reaction time to brain wave period: A reevaluation. *Electroencephalography and Clinical Neurophysiology, 30*, 229–235.

Bonnet, M., Requin, J., & Semjen, A. (1981). Human reflexology and motor preparation. In D. Miller (Ed.), *Exercise and sport science reviews* (pp. 119–157). Philadelphia: Franklin Institute Press.

Bonnet, M., Stelmach, G., & Requin, J. (1982). Specification of direction and extent in motor programming. *Bulletin of the Psychonomics Society, 19*, 31–34.

Bouyer, J., Montaron, M., & Rougeul, A. (1981). Fast frontoparietal rhythms during combined focused attentive behaviour and immobility in cat cortical and thalamic localizations. *Electroencephalography and Clinical Neurophysiology, 51*, 244–252.

Bringuier, V., Frégnac, Y., Baranyi, A., Debanne, D., & Schultz, D. (1997). Synaptic origin and stimulus dependency of neuronal oscillatory activity in the primary visual cortex of the cat. *Journal of Physiology, 500*, 751–774.

Brunia, C. (1999). Neural aspect of anticipatory behavior. *Acta Psychologica, 101*, 213–242.

Burle, B., & Bonnet, M. (1997). Further argument for the existence of a pacemaker in the human information processing system. *Acta Psychologica, 97*, 129–143.

Burle, B., & Bonnet, M. (1999). What's an internal clock for? From temporal information processing to temporal processing of information. *Behavioural Processes, 45*, 59–72.

Burle, B., & Bonnet, M. (2000). High speed memory scanning: A behavioral argument for a serial oscillatory model. *Cognitive Brain Research, 9*, 327–337.

Canu, M., Buser, P., & Rougeul, A. (1994). Relationship between posterior thalamic nucleus unit activity and parietal cortical rhythms (beta) in the waking cat. *Neuroscience, 60*, 679–688.

Caton, R. (1875). The electric currents of the brain. *British Medical Journal, 2*, 278.

Cattaert, D., Semjen, A., & Summers, J. (1999). Simulating a neural crosstalk model for between-hand interference during bimanual circle drawing. *Biological Cybernetics, 81*, 343–358.

Collyer, C., Broadbent, H., & Church, R. (1992). Categorical time perception: Evidence for discrete timing in motor control. *Perception and Psychophysics, 51*, 134–144.

Collyer, C., Broadbent, H., & Church, R. (1994). Preferred rate of tapping and categorical time perception. *Perception and Psychophysics, 55*, 443–453.

Conway, B., Halliday, D., Farmer, S., Shahani, U., Maas, P., Weir, A., & Rosenberg, J. (1995). Synchronization between motor cortex and spinal motoneuronal pool during the performance of a maintained motor task in man. *Journal of Physiology, 489*, 917–924.

Coquery, J. (1994). Processus attentionels. In M. Richelle, J. Requin, & M. Robert (Eds.), *Traité de psychologie expérimentale.* Tome 1 (pp. 219–281). Paris: Presses Universitaires de France.

Crick, F. (1984). Function of the thalamic reticular complex: The searchlight hypothesis. *Proceedings of the National Academy of Sciences USA, 81*, 4586–4590.
Dehaene, S. (1993). Temporal oscillations in human perception. *Psychological Science, 4*, 264–270.
Donders, F. (1868). Over de snelheid van psychische procesen [On the speed of mental processes]. *Archiv für Anatomie und Physiologie, 8*, 657–681.
Fisher, B., & Weber, H. (1993). Express saccades and visual attention. *Behavioral and Brain Sciences, 16*, 553–610.
Frégnac, Y., Bringuier, V., & Baranyi, A. (1994). Oscillatory neuronal activity in visual cortex: A critical re-evaluation. In G. Buzáki, R. Llinás, W. Singer, A. Berthoz, & Y. Christen (Eds.), *Temporal coding in the brain* (pp. 81–102). Berlin: Springer.
Fries, P., Reynolds, J. H., Rorie, A. E., & Desimone, R. (2001). Modulation of oscillatory neuronal synchronization by selective visual attention. *Science, 291*, 1560–1563.
Geissler, H. (1987). The temporal architecture of central information processing: Evidence for a tentative time-quantum model. *Psychological Research, 49*, 99–106.
Geissler, H.-G., & Kompass, R. (2001). Temporal constraints on binding? Evidence from quantal state transitions in perception. *Visual Cognition, 8*, 679–696.
Gibbon, J., Church, R., & Meck, W. (1984). Scalar timing in memory. In J. Gibbon & L. Allan (Eds.), *Annals of the New-York Academy of Science* (Vol. 23, pp. 52–77). New York: New York Academy of Sciences.
Goodman, D., & Kelso, J. (1983). Exploring the functional significance of physiological tremor: A biospectroscopic approach. *Experimental Brain Research, 49*, 419–431.
Gottsdanker, R. (1975). The attaining and maintaining of preparation. In P. M. A. Rabbit & S. Dornic (Eds.), *Attention and Performance V* (pp. 33–49). London: Academic Press.
Gray, C., & Singer, W. (1989). Stimulus specific neuronal oscillation in orientation columns of cat visual cortex. *Proceedings of the National Academy of Sciences USA, 86*, 1698–1702.
Harter, M. (1967). Excitability cycles and cortical scanning: A review of two hypotheses of central intermittency in perception. *Psychological Bulletin, 68*, 47–58.
Harter, M., & White, C. (1968). Periodicity within reaction time distributions and electromyograms. *Quarterly Journal of Experimental Psychology, 20*, 157–166.
Hasbroucq, T., Osman, A., Possamaï, C.-A., Burle, B., Carron, S., Dépy, D., Latour, S., & Mouret, I. (1999). Cortico-spinal, inhibition reflects time but not event preparation: Neural mechanisms of preparation dissociated by transcranial magnetic stimulation. *Acta Psychologica, 101*, 243–266.
Hebb, D. O. (1949). *The organization of behavior*. New York: Wiley.
Herrmann, C., & Knight, R. (2001). Mechanisms of human attention: Event-related potentials and oscillations. *Neuroscience and Biobehavioral Reviews, 25*, 465–476.
Hirsh, I., & Fraisse, P. (1964). Simultanéité et succession de stimuli hétérogènes [Simultaneity and successiveness of heterogeneous stimuli]. *L'Année Psychologique, 64*, 1–19.
Hjorth, B. (1975). An on line transformation of EEG scalp potential into orthogonal source of derivation. *Electroencephalography and Clinical Neurophysiology, 39*, 526–530.
Hohle, R. (1965). Inferred components of reaction times as functions of foreperiod duration. *Journal of Experimental Psychology, 69*, 382–386.
Jefferys, J., Traub, R., & Whittington, M. (1996). Neuronal networks for induced "40 Hz" rhythms. *Trends in Neurosciences, 19*, 202–208.
Jensen, O., & Lisman, J. (1998). An oscillatory short-term memory buffer model can account for data on the Sternberg task. *Journal of Neuroscience, 18*, 10688–10699.

Kelso, J., Holt, K., Rubin, P., & Kugler, P. (1981). Patterns of human interlimb coordination emerge from the properties of non-linear, limit cycle oscillatory processes: theory and data. *Journal of Motor Behavior, 13*, 226–261.

Kirschfeld, K., Feiler, R., & Wolf-Oberhollenzer, F. (1996). Cortical oscillations and the origin of express saccades. *Proceedings of the Royal Society of London B, 263*, 459–468.

Kristofferson, A. B. (1967). Attention and psychological time. *Acta Psychologica, 27*, 93–100.

Kristofferson, A. B. (1976). Low-variance stimulus-response latencies: Deterministic internal delays? *Perception and Psychophysics, 20*, 89–100.

Kristofferson, A. B. (1980). A quantal step function in duration discrimination. *Perception and Psychophysics, 27*, 300–306.

Kristofferson, A. B. (1984). Quantal and deterministic timing in human duration discrimination. In J. Gibbon & L. Allan (Eds.), *Annals of the New York Academy of Science.* Volume 23 (pp. 3–15). New York: New York Academy of Science.

LaBerge, D. (1995). *Attentional processing.* Cambridge, MA: Harvard University Press.

Lakie, M., & Combes, N. (2000). There is no simple temporal relationship between the initiation of rapid reactive hand movements and the phase of an enhanced physiological tremor in man. *Journal of Physiology, 523*, 515–522.

Latour, P. (1967). Evidence of internal clocks in the human operator. *Acta Psychologica, 27*, 341–348.

Lesèvre, N., & Rémond, A. (1967). Variations de la réponse visuelle moyenne en fonction de la phase de l'alpha [Variations of the average visual response as a function of the alpha phase]. *Revue Neurologique, 117*, 215–216.

Lisman, J. (1997). Bursts as a unit of neural information: Making unreliable synapses reliable. *Trends in Neurosciences, 20*, 38–43.

Lisman, J., & Idiart, M. (1995). Storage of 7 ± 2 short-term memories in oscillatory subcycles. *Science, 267*, 1512–1515.

Llinás, R., Grace, A., & Yarom, Y. (1991). In vitro neurons in mammalian cortical layer 4 exhibit intrinsic oscillatory activity in the 10- to 50-Hz frequency range. *Proceedings of the National Academy of Sciences USA, 88*, 897–901.

Llinás, R., Ribary, U., Joliot, M., & Wang, X. (1994). Content and context in temporal thalamocortical binding. In G. Buzáki, R. Llinás, W. Singer, A. Berthoz, & Y. Christen (Eds.), *Temporal coding in the brain* (pp. 251–272). Berlin: Springer.

Luce, R. (1986). *Response times: Their roles in infering mental organization.* New York: Oxford University Press.

Lytton, W., & Sejnowsky, T. (1991). Simulations of cortical pyramidal neurons synchronized by inhibitory interneurons. *Journal of Neurophysiology, 66*, 1059–1078.

MacKay, W. (1997). Synchronized neuronal oscillations and their role in motor processes. *Trends in Cognitive Sciences, 1*, 176–183.

McAuley, J., Farmer, S., Rothwell, J., & Marsden, C. (1999). Common 3 and 10 Hz oscillations modulate human eye and finger movements while they simultaneously track a visual target. *Journal of Physiology, 515 (3)*, 905–917.

McKay, D. (1983). On-line source density computation with a minimum of electrodes. *Electroencephalography and Clinical Neurophysiology, 56*, 696–698.

Miller, G. A. (1956). The magical number seven, plus minus two: Some limits on our capacity for processing information. *Psychological Review, 63*, 81–97.

Müller, M., & Gruber, T. (2001). Induced gamma-band responses in the human EEG are related to attentional information processing. *Visual Cognition, 8*, 579–592.

Munk, M., Roelfsema, P., König, P., Engel, A., & Singer, W. (1996). Role of the reticular activation in the modulation of intracortical synchronization. *Science, 272*, 271–274.

Murthy, V., & Fetz, E. (1996). Synchronization of neurons during local field potential oscillations in sensorimotor cortex of awake monkey. *Journal of Neurophysiology, 76*, 3968–3982.

Nashmi, R., Mendoça, A., & MacKay, W. (1994). EEG rhythms of the sensorimotor region during hand movements. *Electroencephalography and Clinical Neurophysiology, 91*, 456–467.

Nuñez, A., Amzica, F., & Steriade, M. (1992). Intrinsic and synaptically generated delta (1–14 Hz) rhythms in dorsal lateral geniculate neurons and their modulation by light-induced fast (30–70 Hz) events. *Neuroscience, 51*, 269–284.

Pinault, D., & Deschênes, M. (1992). Voltage-dependent 40-Hz oscillations in rat reticular thalamic neurons in vivo. *Neuroscience, 51*, 245–258.

Pöppel, E. (1996). Reconstruction of subjective time on the basis of hierarchically organized processing system. In M. Pastor & J. Artieda (Eds.), *Time, internal clocks and movement* (pp. 165–185). Amsterdam, Netherlands: North-Holland/Elsevier Science Publishers.

Posner, M. (1978). *Chronometric explorations of mind.* Hillsdale, N.J.: Erlbaum.

Proctor, R., Reeve, T., &Weeks, D. (1990). A triphasic approach to the acquisition of response-selection skill. In G. Bower (Ed.), *The psychology of learning and motivation* (pp. 207–240). New York: Academic Press.

Purves, D., Paydarfar, J., & Andrews, T. (1996). The wagon wheel illusion in movies and reality. *Proceedings of the National Academy of Sciences USA, 93*, 3693–3697.

Requin, J., Brener, J., & Ring, C. (1991). Preparation for action. In J. Jennings & M. Coles (Eds.), *Handbook of cognitive psychophysiology: Central and autonomic nervous system approaches* (pp. 357–448). New York: Wiley.

Riehle, A., Grün, S., Diesmann, M., & Aertsen, A. (1997). Spike synchronization and rate modulation differentially involved in motor cortical function. *Science, 278*, 1950–1953.

Roelfsema, P. R., Engel, A., König, P., & Singer, W. (1997). Visuomotor integration is associated with zero time-lag synchronization among cortical areas. *Nature, 385*, 157–161.

Rosembaum, D. (1980). Human movement initiation: specification of arm direction and extent. *Journal of Experimental Psychology: General, 109*, 444–474.

Sanders, A. (1998). *Elements of human performance: reaction processes and attention in human skill.* Mahwah: Erlbaum.

Sanes, J., & Donohogue, J. (1993). Oscillations in local field potentials of the primate motor cortex during voluntary movement. *Proceedings of the National Academy of Sciences USA, 90*, 4470–4474.

Schieppati, M. (1987). The Hoffman reflex: A means of assessing spinal reflex excitability and descending control in man. *Progress in Neurobiology, 28*, 345–376.

Schmidt, M., & Kristofferson, A. B. (1963). Discrimination of successiveness: A test of a model of attention. *Science, 139*, 112–113.

Shallice, T. (1964). The detection of change and the perceptual moment hypothesis. *The British Journal of Statistical Psychology, 17*, 113–135.

Sheer, D. (1970). Electrophysiological correlates in memory consolidation. In G. Ungar (Ed.), *Molecular mechanisms in memory and learning* (pp. 177–211). NewYork: Plenum Press.

Sheer, D. (1976). Focused arousal and 40Hz EEG. In R. Knight & D. Bakker (Eds.), *The neuropsychology of learning disorders* (pp. 71–87). Baltimore, MD: University Park Press.

Sherrington, C. (1947). *The integrative action of the nervous system.* Second edition. New Haven, CT: Yale University Press.

Singer, W. (1993). Synchronization of cortical activity and its putative role in information processing and learning. *Annual Review of Physiology, 55*, 349–374.
Steriade, M. (1993). Central core modulation of spontaneous oscillations and sensory transmission in thalamocortical systems. *Current Opinion in Neurobiology, 3*, 619–625.
Steriade, M., & Amzica, F. (1996). Intracortical and corticothalamic coherency of fast spontaneous oscillations. *Proceedings of the National Academy of Sciences USA, 93*, 2533–2538.
Steriade, M., Curró-Dossi, R., Paré, D., & Oakson, G. (1991). Fast oscillations (20–40 Hz) in thalamocortical systems and their potentiation by mesopontine cholinergic nuclei in the cat. *Proceedings of the National Academy of Sciences USA, 88*, 4396–4400.
Sternberg, S. (1966). High speed scanning in human memory. *Science, 153*, 652–654.
Sternberg, S. (1969a). The discovery of processing stages: Extension of Donder's method. *Acta Psychologica, 30*, 276–315.
Sternberg, S. (1969b). Memory scanning: mental processes revealed by reaction time experiment. *American Scientist, 57*, 421–457.
Sternberg, S. (1998). Discovering mental processing stages: The method of additive factors. In D. Scarborough & S. Sternberg (Eds.), *An invitation to cognitive science, methods, models, and conceptual issues.* Volume. 4 (pp. 703–863). Cambridge: MIT Press.
Stroud, J. (1955). The fine structure of psychological time. In H. Quastler (Ed.), *Information theory in psychology*. Glencoe, IL: Free Press.
Surwillo, W. (1961). Frequency of the α rhythm, reaction time and age. *Nature, 191*, 823–824.
Surwillo, W. (1963). The relation of simple response time to brain wave frequency and the effects of age. *Electroencephalography and Clinical Neurophysiology, 16*, 510–514.
Surwillo, W. (1964). The relation of decision time to brain wave frequency and to age. *Electroencephalography and Clinical Neurophysiology, 16*, 510–514.
Timofeev, Y., & Steriade, M. (1997). Fast (mainly 30–100 Hz) oscillations in the cat cerebellothalamic pathway and their synchronization with cortical potentials. *Journal of Physiology, 504*, 153–168.
Titinen, H., Sinkkonen, J., Reinikainen, K., Alho, K., Lavikainen, V., & Naätänen, R. (1993). Selective attention enhances the auditory 40 Hz transient response in humans. *Nature, 364*, 59–60.
Traub, R., Whittington, M., Stanford, I., & Jefferys, J. (1996). A mechanism for generation of long-range synchronous fast oscillations in the cortex. *Nature, 383*, 621–624.
Travis, L. (1929). The relation of voluntary movement to tremors. *Journal of Experimental Psychology, 12*, 515–524.
Travis, L., & Cofer, C. (1937). The temporal relationship between brain potentials and certain neuromuscular rhythms. *Journal of Experimental Psychology, 20*, 565–569.
Treisman, M. (1963). Temporal discrimination and the indifference interval: Implications for a model of the "internal clock". *Psychological Monographs, 77*, 1–31.
Treisman, M. (1984). Temporal rhythms and cerebral rhythms. In J. Gibbon & L. Allan (Eds.), *Annals of the New-York Academy of Science.* Volume 423 (pp. 542–565). New York: New York Academy of Sciences.
Treisman, M., Faulkner, A., & Naish, P. (1992). On the relation between time perception and the timing of motor action: Evidence for a temporal oscillator controlling the timing of movement. *Quarterly Journal of Experimental Psychology, 45A*, 235–263.
Treisman, M., Faulkner, A., Naish, P., & Brogan, D. (1990). The internal clock: Evidence for a temporal oscillator underlying time perception with some estimates of its characteristic frequency. *Perception, 19*, 705–743.

Treisman, M., Faulkner, A., Naish, P., & MacCrone, J. (1994). The internal clock: Electroencephalographic evidence for oscillatory processes underlying time perception. *Quarterly Journal of Experimental Psychology, 47A*, 241–289.

Vallbo, A., & Wessberg, J. (1993). Organization of motor output in slow finger movements in man. *Journal of Physiology, 469*, 673–691.

Varela, F., Toro, A., John, E., & Schwartz, E. (1981). Perceptual framing and cortical alpha rhythm. *Neuropsychologia, 19* (5), 675–686.

Vidal, F., Bonnet, M., & Macar, F. (1995). Programming the duration of a motor sequence: role of the primary and supplementary motor areas in man. *Experimental Brain Research, 106*, 339–350.

Vorberg, D., & Hambuch, R. (1984). Timing of two-handed performance. In J. Gibbon & L. Allan (Eds.), *Annals of the New York Academy of Science.* Volume 423 (pp. 390–406). New York: New York Academy of Sciences.

Vorberg, D., & Schwartz, W. (1987). Oscillatory mechanisms in human reaction times? *Naturwissenschaften, 74*, 446–447.

Vroon, P. (1974). Is there a time quantum in duration experience? *American Journal of Experimental Psychology, 87*, 237–245.

Wiener, N. (1958). Time and the science of organisation. *Scientia, 93*, 199–205.

Wing, A. M., & Kristofferson, A. B. (1973a). Response delays and the timing of discrete motor responses. *Perception and Psychophysics, 14*, 5–12.

Wing, A. M., & Kristofferson, A. B. (1973b). The timing of interresponse intervals. *Perception and Psychophysics, 13*, 455–460.

Woodruff, D. (1975). Relationships among EEG alpha frequency, reaction time, and age: A biofeedback study. *Psychophysiology, 12*, 673–681.

Subject Index

G

I

J

K

L

M

Author Index

E

F

G

H

I

J

K

L

M

N

O

P

R

S

T